Van De Graaff's Photographic Atlas

for the

Anatomy and Physiology Laboratory

EIGHTH EDITION

David A. Morton

University of Utah

John L. Crawley

MORTON PUBLISHING

925 W. Kenyon Avenue, Unit 12
Englewood, CO 80110

www.morton-pub.com

Book Team

Publisher	Douglas N. Morton
President	David M. Ferguson
Acquisitions Editor	Marta R. Martins
Project Editor	Rayna S. Bailey
Editorial Assistant	Sarah D. Thomas
Production Team	Joanne Saliger, Will Kelley, and Sarah Bailey
Design and Layout	John Crawley
Illustrations	Imagineering Media Services, Inc.

To Kent, a teacher, mentor, colleague, and friend.

Printed in the United States of America

10 9 8 7 6 5 4 3 2

ISBN-10: 1-61731-277-0

ISBN-13: 978-1-61731-277-9

Library of Congress Control Number: 2015907555

Table of Contents

A Photographic Atlas for the Anatomy and Physiology Laboratory

Preface

Human anatomy is the scientific discipline that investigates the structure of the body and human physiology is the scientific discipline that investigates how body structures function. These subjects may be taught independent of each other in separate courses, or they may be taught together in integrated anatomy and physiology courses. Regardless of whether or not anatomy is taught independently from physiology or if the two disciplines are integrated as a single course, it is necessary for a student to have a conceptualized visualization of body structure and a knowledge of its basic descriptive anatomical terminology in order to understand how the body functions.

A Photographic Atlas for the Anatomy and Physiology Laboratory is designed for all students taking separate or integrated courses in human anatomy and physiology. This atlas can accompany and will augment any human anatomy, human physiology, or combined human anatomy and physiology textbook. It is designed to be of particular value to students in a laboratory situation and could either accompany a laboratory manual or in certain courses, serve as the laboratory manual.

Anatomy and physiology are visually oriented sciences. Great care has gone into the preparation of this photographic atlas to provide students with a complete set of photographs for each of the human body systems. Human cadavers have been carefully dissected and photographs taken that clearly depict each of the principal organs from each of the body systems. Cat dissection, fetal pig dissection, and rat dissection are also included for those students who have the opportunity to do similar dissection as part of their laboratory requirement. In addition, photographs of a sheep heart dissection are also included.

A visual balance is achieved in this atlas between the various levels available to observe the structure of the body. Microscopic anatomy is presented by photomicrographs at the light microscope level and electron micrography from scanning and transmission electron microscopy. Carefully selected photographs are used throughout the atlas to provide a balanced perspective of the gross anatomy. At the request of several professors who used previous editions of the atlas, the muscular and circulatory sections have been expanded and improved with new photographs, illustrations, and tables. The section on articulations has been improved with the inclusion of photographs of joint dissections. Selected X-rays, CT scans, and MR images depict structures from living persons and thus provide an applied dimension to the atlas. Great care has been taken to construct completely labeled, informative figures that are depicted clearly and accurately. The terminology used in this atlas are those that are approved and recommended by the Basle Nomina Anatomica (BNA).

Preface to Eighth Edition

New editions are desirable for authors because it presents an opportunity to improve upon a successful product. Revision, such as is presented in the eighth edition of *A Photographic Atlas for the Human Anatomy and Physiology Laboratory*, requires an inordinate amount of planning, organization, and work. As authors we have the opportunity and obligation to listen to the critiques and suggestions from students and faculty who have used this atlas. This constructive input has resulted in a product that continues to improved. We appreciate those who have taken the time to provide suggestions and indicate corrections.

One of the objectives in preparing this atlas was to create an inviting pedagogy. The page layout has been improved by careful selection of photographs, and when necessary, provide accompanying line art. Each image in this atlas has been carefully evaluated for its quality, effectiveness, and accuracy. Black backgrounds for the depicted specimens enhance the clarity of the images.

Acknowledgments

Many individuals contributed to the preparation of the eighth edition of *A Photographic Atlas for the Human Anatomy and Physiology Laboratory*. We are especially appreciative of Chris Steadman, Aaron Bera and Steven Taylor who helped conduct the tedious and meticulous dissections of the cadavers. They were enjoyable to work with and were conscientious in meeting the dissection schedule. We are also grateful for Dr. Robert Seegmiller of Brigham Young University for his help in acquiring specimens.

It is gratifying to have professors and health-care professionals interested in the success of *A Photographic Atlas for the Human Anatomy and Physiology Laboratory*. There are several that were helpful in the development of this atlas. They share our enthusiasm of its value for students of anatomy and physiology. We are especially appreciative of Kyle M. Van De Graaff, M.D. and William B. Winborn, Ph.D. for their efforts and generosity in providing the choice photomicrographs used in this atlas. The radiographs, CT scans, and MR images were made possible through the generosity of Gary M. Watts, M.D. and the Department of Radiology at Utah Valley Regional Medical Center. Kerry Peterson for the use of his dissections. We thank Jake Christiansen, James Barrett and Austen Slade for their specimen dissections. Others who aided in specimen dissections were Nathan A. Jacobson, D.O., R. Richard Rasmussen, M.D., and Sandra E. Sephton, Ph.D. We appreciate the talents of Imagineering who rendered the line art throughout the atlas. Many users and reviewers of the previous editions of this atlas provided suggestions for its improvement. We are especially appreciative of Michael J. Shively, D.V.M. for his numerous comments and helpful suggestions. Special thanks to Jill E. Feinstein, Richland Community College, Ruby E. Fogg, Manchester Community College, NH, Amanda L. Posey, Lamar University, Amanda Rosenzweig, Delgado Community College, Patricia Visser, Jackson College, for their help in reviewing this atlas. We are indebted to Douglas Morton and the personnel at Morton Publishing Company for the opportunity, encouragement, and support to complete this project.

Body Organization

Chapter 1

Anatomy is the study of body structures. An example of an anatomical study is learning about the structure of the heart—the chambers, valves, and vessels that serve the heart muscle. **Physiology** is the study of body function. An example of a physiological study is learning what causes the heart muscles to contract—the sequence of blood flow through the heart and what causes blood pressure. The anatomy (structure) and the physiology (function) of any part of the body are always related, or in other words, structure determines function.

Most of the physiological processes within the body act to maintain **homeostasis**. Simply defined, homeostasis is maintaining nearly consistent internal conditions within the body despite changing conditions in the external environment. For example, one area of your brain acts as a thermostat to keep your body temperature near 37°C (98.6°F). Being too warm causes you to sweat and cool the body, while being cold causes you to shiver and warm the body. Maintaining overall body homeostasis is achieved through many interacting physiological processes involving all levels of body organization, and is absolutely necessary for survival.

Structural and functional levels of organization exist in the body, and each of its parts contributes to the total organism. In the study of human anatomy and physiology, the following levels of body organization are generally recognized—the molecular level, the cellular level, the tissue level, the organ level, the system level, and the organismic level (Fig. 1.1).

Cells are microscopic and are the smallest living part of all organisms. **Tissues** are of groups of similar cells that perform specific functions. An **organ** is an aggregate of two or more tissues integrated to perform a particular function. The **systems** of the body consist of various body organs that have similar or related functions. All the systems of the body are interrelated and function together constituting the **organism**.

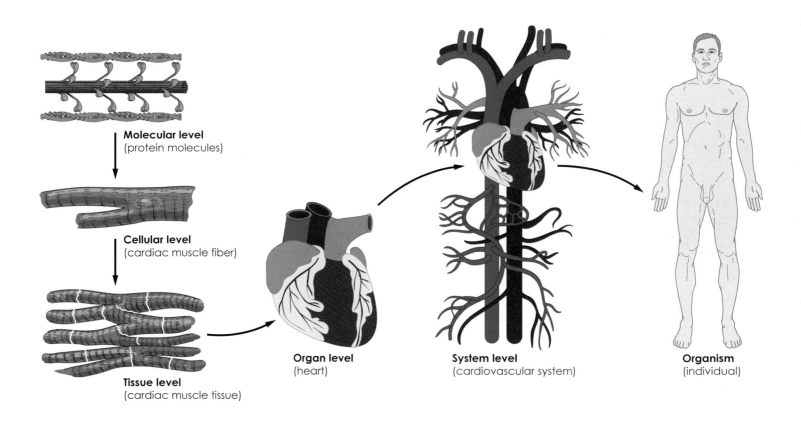

Molecular level
(protein molecules)

Cellular level
(cardiac muscle fiber)

Tissue level
(cardiac muscle tissue)

Organ level
(heart)

System level
(cardiovascular system)

Organism
(individual)

Figure 1.1 The levels of structural organization and complexity within the body.

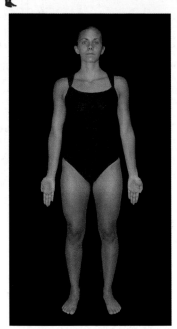

Figure 1.2 The anatomical position provides a basis of reference for describing the relationship of one body part to another. In the anatomical position, the person is standing, the feet are parallel, the eyes are directed forward, and the arms are to the sides with the palms turned forward and the fingers are pointed straight down.

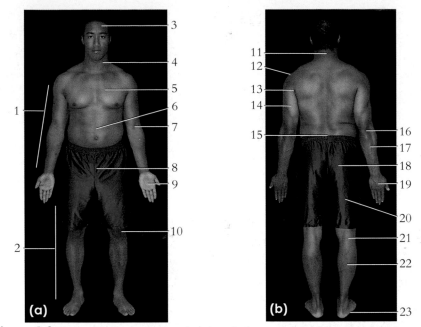

Figure 1.3 The major body parts and regions in humans (bipedal vertebrate). (a) An anterior view and (b) a posterior view.

1. Upper extremity
2. Lower extremity
3. Head
4. Neck, anterior aspect
5. Thorax (chest)
6. Abdomen
7. Cubital fossa
8. Pubic region
9. Palmar region (palm)
10. Patellar region (patella)
11. Cervical region
12. Shoulder
13. Axilla (armpit)
14. Brachium (upper arm)
15. Lumbar region
16. Elbow
17. Antebrachium (forearm)
18. Gluteal region (buttock)
19. Dorsum of hand
20. Thigh
21. Popliteal fossa
22. Calf
23. Plantar surface (sole)

Table 1.1 Directional terminology for describing human body structures.

Term	Definition	Example
Superior (cephalic, cranial)	Toward the top of the head	The neck is superior to the thorax.
Inferior (caudal)	Away from the top of the head	The pubic region is inferior to the abdomen.
Anterior (ventral)	Toward the front of the body	The eyes, nose, and mouth are on the anterior side of the body.
Posterior (dorsal)	Toward the back of the body	The spinal cord extends down the posterior side of the body.
Lateral	Toward the side of the body	The arms are on the lateral sides of the body.
Medial	Toward the median plane of the body	The heart is medial to the lungs.
Superficial (external)	Toward the surface of the body	The skin is superficial to the muscles.
Deep (internal)	Away from the surface of the body	The heart is positioned deep within the thoracic cavity.
Parietal	Reference to the body wall of the trunk (thorax and abdomen)	The parietal peritoneum is the membrane lining the abdominal cavity.
Visceral	Reference to internal organs of trunk	The stomach is covered by a thin membrane called the visceral peritoneum.

Table 1.2 Directional terminology for describing quadrupedal body structures.

Term	Definition	Example
Cranial	Toward the head	The neck is cranial to the thorax.
Caudal	Toward the tail	The pubic region is caudal to the abdomen.
Dorsal	The back (equivalent to posterior when referring to the human body)	The shoulder blade is dorsal to the rib cage.
Ventral	The belly side (equivalent to anterior when referring to the human body)	The navel is on the ventral surface of the trunk.

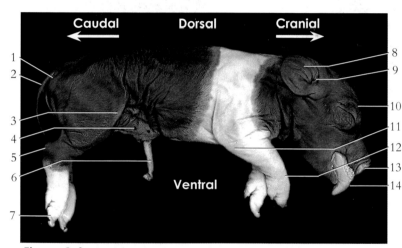

Figure 1.4 The directional terminology and superficial structures in a fetal pig (quadrupedal vertebrate).

1. Anus
2. Tail
3. Knee
4. Teat
5. Ankle
6. Umbilical cord
7. Hoof
8. Auricle (pinna)
9. External auditory canal
10. Superior palpebra (superior eyelid)
11. Elbow
12. Wrist
13. Naris (nostril)
14. Tongue

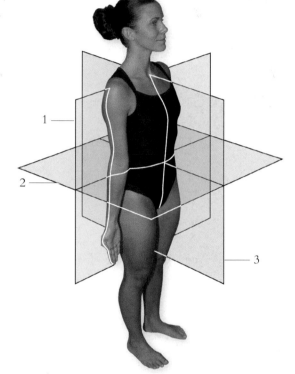

Figure 1.5 The planes of reference in a human (bipedal vertebrate).
1. Coronal plane (frontal plane)
2. Transverse plane (cross-sectional plane)
3. Sagittal plane

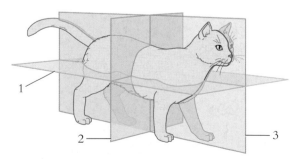

Figure 1.6 The directional terminology and superficial structures in a cat (quadrupedal vertebrate).

1. Thigh
2. Tail
3. Auricle (pinna)
4. Superior palpebra (superior eyelid)
5. Bridge of nose
6. Naris (nostril)
7. Vibrissae
8. Brachium
9. Manus (front foot)
10. Claw
11. Antebrachium

Figure 1.7 The planes of reference in a cat (quadrupedal vertebrate).
1. Coronal plane (frontal plane)
2. Transverse plane (cross-sectional plane)
3. Sagittal plane

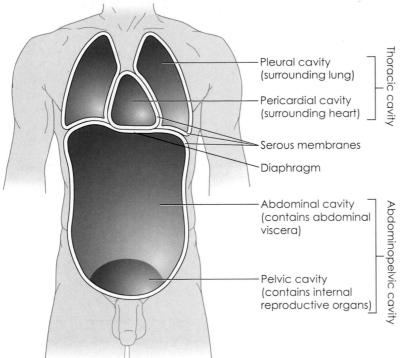

Pleural cavity
(surrounding lung)

Pericardial cavity
(surrounding heart)

Thoracic cavity

Serous membranes

Diaphragm

Abdominal cavity
(contains abdominal
viscera)

Pelvic cavity
(contains internal
reproductive organs)

Abdominopelvic cavity

Figure 1.8 An anterior view of the body cavities of the trunk.

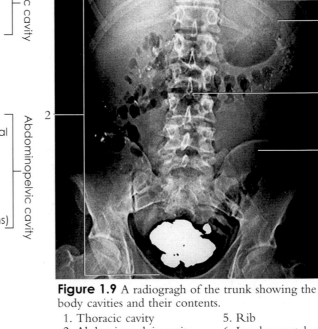

Figure 1.9 A radiogragh of the trunk showing the body cavities and their contents.

1. Thoracic cavity
2. Abdominopelvic cavity
3. Heart
4. Diaphragm
5. Rib
6. Lumbar vertebra
7. Ilium

Cranial cavity
(contains brain)

Vertebral cavity
(contains spinal cord)

Posterior (dorsal) cavity

Thoracic cavity

Serous membranes

Abdominopelvic
cavity

Anterior (ventral) cavity

Figure 1.10 A midsagittal view of the body cavities.

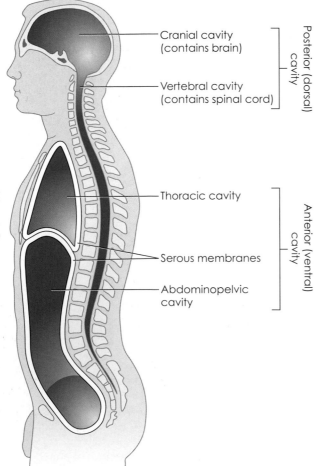

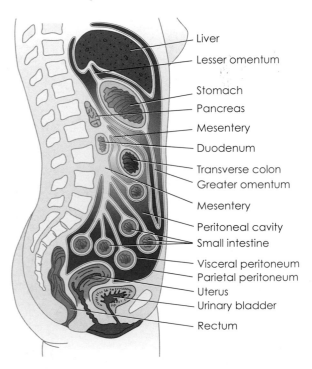

Liver

Lesser omentum

Stomach

Pancreas

Mesentery

Duodenum

Transverse colon

Greater omentum

Mesentery

Peritoneal cavity

Small intestine

Visceral peritoneum

Parietal peritoneum

Uterus

Urinary bladder

Rectum

Figure 1.11 A midsagittal view of the organs of the abdominopelvic cavity and their supporting membranes.

Figure 1.12 The human male.
(a) Anterior view
(b) Posterior view
 1. Facial region
 2. Cranial region
 3. Posterior neck
 4. Anterior neck
 5. Shoulder
 6. Thorax
 7. Nipple
 8. Brachium
 9. Elbow
10. Cubital fossa
11. Abdomen
12. Umbilicus (navel)
13. Antebrachium
14. Wrist (carpus)
15. Hand
16. Natal (gluteal) cleft
17. Fold of buttock (gluteal fold)
18. External genitalia
19. Thigh
20. Patella
21. Popliteal fossa
22. Leg
23. Ankle
24. Foot

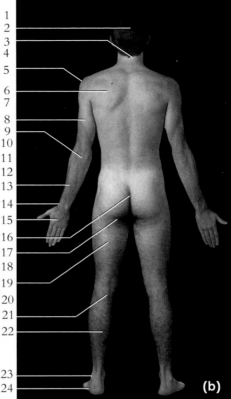

(a) (b)

Figure 1.13 The human female.
(a) Anterior view
(b) Posterior view
 1. Facial region
 2. Cranial region
 3. Posterior neck
 4. Anterior neck
 5. Shoulder
 6. Thorax
 7. Breast
 8. Nipple
 9. Brachium
10. Cubital fossa
11. Elbow
12. Abdomen
13. Antebrachium
14. Iliac crest
15. Umbilicus (navel)
16. Wrist (carpus)
17. Hand
18. Natal (gluteal) cleft
19. Fold of buttock (gluteal fold)
20. Mons pubis
21. Thigh
22. Popliteal fossa
23. Patella
24. Leg
25. Ankle (tarsus)
26. Foot

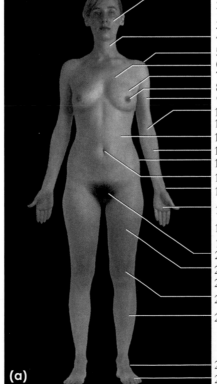

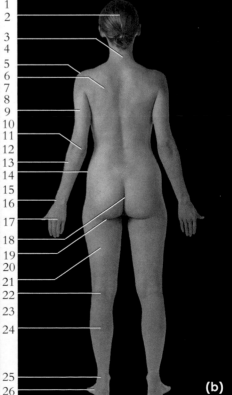

(a) (b)

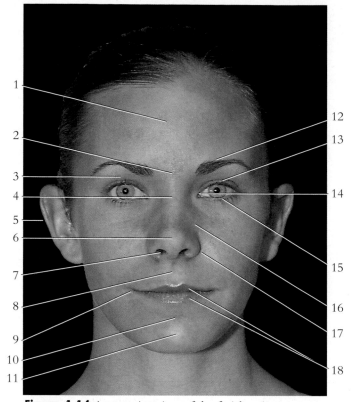

Figure 1.14 An anterior view of the facial region.

1. Forehead
2. Root of nose (glabella)
3. Superior palpebral sulcus
4. Bridge of nose
5. Auricle (pinna)
6. Apex of nose
7. Nostril
8. Philtrum
9. Corner of mouth
10. Mentolabial sulcus
11. Chin (mentalis)
12. Eyebrow
13. Eyelashes of upper eyelid
14. Lacrimal caruncle
15. Eyelashes of lower eyelid
16. Nasofacial angle
17. Alar nasal sulcus
18. Lips

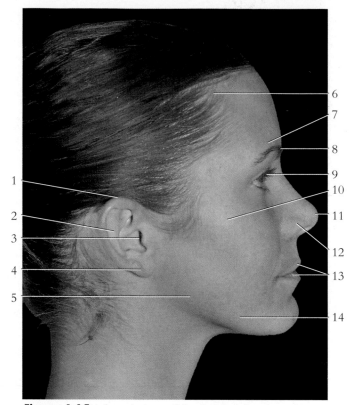

Figure 1.15 A lateral view of the facial region.

1. Helix of auricle
2. Antihelix
3. External auditory canal
4. Earlobe
5. Angle of mandible
6. Hair line
7. Superciliary ridge
8. Eyebrow
9. Eyelashes
10. Zygomatic arch
11. Apex of nose
12. Ala nasi
13. Lips
14. Body of mandible

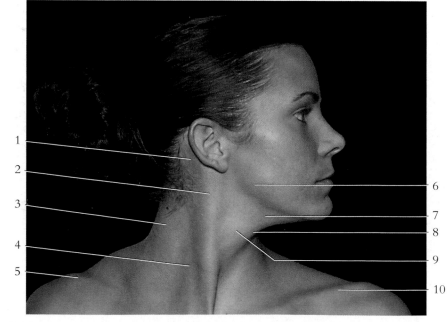

Figure 1.16 An anterolateral view of the neck. (m. = muscle)

1. Mastoid process
2. Sternocleidomastoid m.
3. Trapezius m.
4. Posterior triangle of neck
5. Acromion of scapula
6. Angle of mandible
7. Hyoid bone
8. Thyroid cartilage of larynx
9. Anterior triangle of neck
10. Clavicle

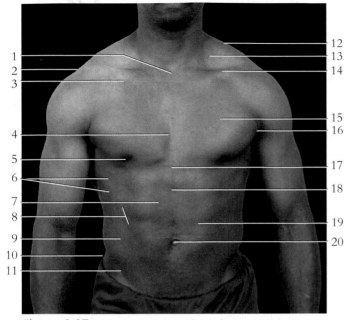

Figure 1.17 An anterior view of the thorax and abdomen (m.=muscle, mm.=muscles).

1. Jugular notch
2. Acromion of scapula
3. Clavipectoral triangle
4. Sternum
5. Nipple
6. Serratus anterior m.
7. Rectus abdominis m.
8. Linea semilunaris
9. External abdominal oblique m.
10. Iliac crest
11. Inguinal ligament
12. Trapezius m.
13. Supraclavicular fossa
14. Clavicle
15. Pectoralis major m.
16. Anterior axillary fold
17. Xiphoid process
18. Linea alba
19. Tendinous inscription
20. Umbilicus

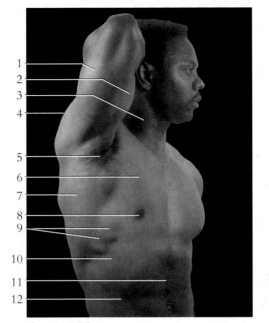

Figure 1.18 An anterolateral view of the thorax, abdomen, and axilla.

1. Triceps brachii m.
2. Biceps brachii m.
3. Sternocleidomastoid m.
4. Deltoid m.
5. Axilla
6. Pectoralis major m.
7. Latissimus dorsi m.
8. Nipple
9. Serratus anterior m.
10. Intercostal m.
11. Linea alba
12. External abdominal oblique m.

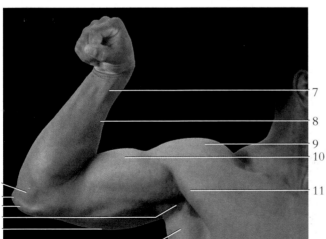

Figure 1.19 The right shoulder, axilla, and upper extremity.

1. Medial epicondyle of humerus
2. Olecranon of ulna
3. Ulnar sulcus
4. Axilla
5. Triceps brachii m.
6. Latissimus dorsi m. (posterior axillary fold)
7. Tendon of flexor carpi radialis m.
8. Brachioradialis m.
9. Deltoid m.
10. Biceps brachii m.
11. Pectoralis major m. (anterior axillary fold)

Figure 1.20 A posterior view of the thorax.

1. Deltoid m.
2. Trapezius m.
3. Infraspinatus m.
4. Triangle of auscultation
5. Inferior angle of scapula
6. Latissimus dorsi m.
7. Median furrow over vertebral column
8. Erector spinae m.

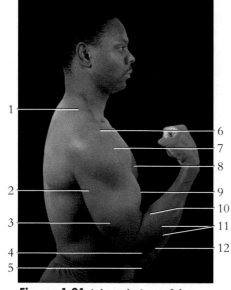

Figure 1.21 A lateral view of the right shoulder and upper extremity.
1. Trapezius m.
2. Long head of triceps brachii m.
3. Lateral head of triceps brachii m.
4. Lateral epicondyle of humerus
5. Olecranon of ulna
6. Acromion of scapula
7. Deltoid m.
8. Pectoralis major m.
9. Biceps brachii m.
10. Brachioradialis m.
11. Extensor carpi radialis longus and brevis mm.
12. Extensor digitorum m.

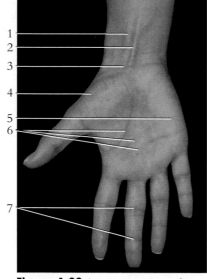

Figure 1.22 An anterior view of the right hand.
1. Tendon of flexor carpi radialis m.
2. Tendon of palmaris longus m.
3. Flexion crease on wrist
4. Thenar eminence
5. Hypothenar eminence
6. Flexion creases on palm of hand
7. Flexion creases on third digit

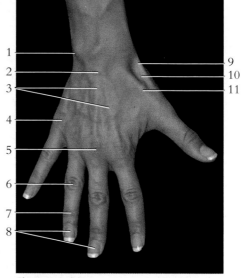

Figure 1.23 A posterior view of the right hand.
1. Styloid process of ulna
2. Position of extensor retinaculum
3. Tendons of extensor digitorum m.
4. Tendon of extensor digiti minimi m.
5. Metacarpophalangeal joint
6. Proximal interphalangeal joint
7. Distal interphalangeal joint
8. Nails
9. Tendon of extensor pollicis brevis m.
10. Anatomical snuffbox
11. Tendon of extensor pollicis longus m.

Figure 1.24 An anterior view of the right upper extremity.
1. Cephalic vein
2. Biceps brachii m.
3. Cubital fossa
4. Brachioradialis m.
5. Cephalic vein
6. Site for palpation of radial artery
7. Tendon of flexor carpi radialis m.
8. Tendon of palmaris longus m.
9. Thenar eminence
10. Metacarpophalangeal joint of thumb
11. Site of palpation of brachial artery
12. Basilic vein
13. Median cubital vein
14. Ulnar vein
15. Median antebrachial vein
16. Tendon of superficial digital flexor m.
17. Styloid process of ulna
18. Hypothenar eminence

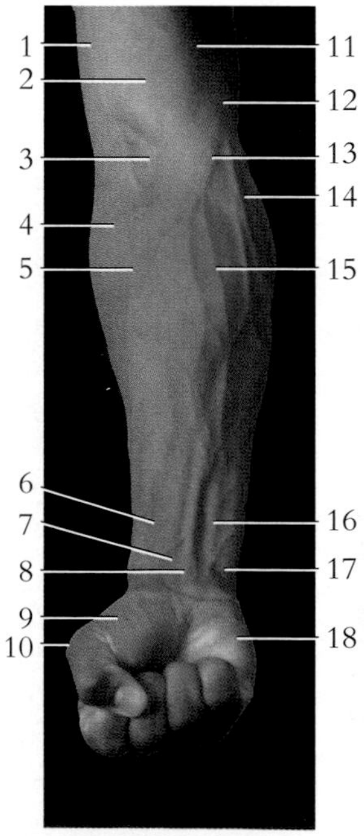

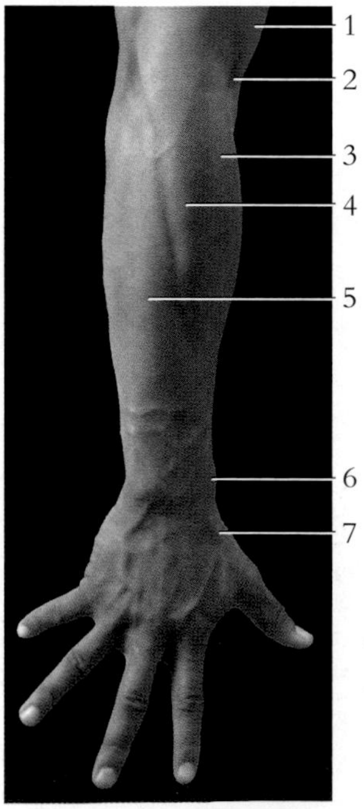

Figure 1.25 A posterior view of the right upper extremity.
1. Biceps brachii m.
2. Cubital fossa
3. Brachioradialis m.
4. Extensor carpi radialis longus m.
5. Extensor carpi ulnaris m.
6. Styloid process of radius
7. Tendon of extensor pollicus longus m.

Figure 1.26 An anterior view of the right thigh.
1. Site of femoral triangle
2. Quadriceps femoris group of muscles
2a. Rectus femoris m.
2b. Vastus lateralis m.
2c. Vastus medialis m.
2d. Tendon of quadriceps femoris m.
3. Patella
4. Patellar ligament
5. Adductor group of muscles
5a. Sartorius m.

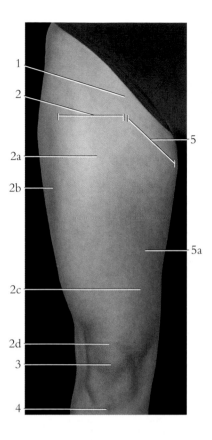

Figure 1.27 A medial view of the right thigh.
1. Adductor magnus m.
2. Gracilis m.
3. Rectus femoris m.
4. Sartorius m.
5. Vastus medialis m.
6. Semimembranosus m.
7. Semitendinosus m.
8. Patella

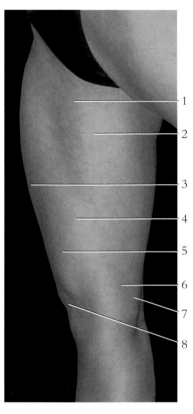

Figure 1.28 A posterior view of the right thigh.
1. Gluteus maximus m.
2. Fold of buttock (gluteal fold)
3. Hamstring group of muscles
3a. Long head of biceps femoris m.
3b. Semitendinosus m.
4. Gracilis m.
5. Popliteal fossa
6. Lateral head of gastrocnemius m.
7. Vastus lateralis m.

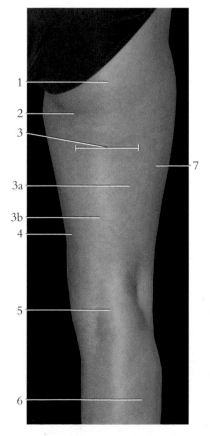
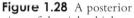

Figure 1.29 A lateral view of the right thigh.
1. Gluteus maximus m.
2. Tensor fasciae latae m.
3. Vastus lateralis m.
4. Rectus femoris m.
5. Biceps femoris m.
6. Iliotibial tract
7. Patella
8. Lateral epicondyle of femur
9. Popliteal fossa

Figure 1.30 An anterior view of the right leg and foot.
1. Patella
2. Patellar ligament
3. Tibialis anterior m.
4. Lateral malleolus of fibula
5. Medial malleolus of tibia
6. Site for palpation of dorsal pedis artery
7. Tendons of extensor digitorum longus m.
8. Tendon of extensor hallucis longus m.

Figure 1.31 A medial view of the right leg and foot.
1. Tibia
2. Medial head of gastrocnemius m.
3. Soleus m.
4. Tendo calcaneus
5. Medial malleolus of tibia
6. Calcaneus
7. Abductor hallucis m.
8. Longitudinal arch
9. Tendon of extensor hallucis longus m.
10. Head of first metatarsal bone

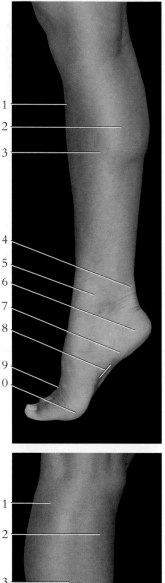

Figure 1.32 A posterior view of the right leg and foot.
1. Popliteal fossa
2. Lateral head of gastrocnemius m.
3. Medial head of gastrocnemius m.
4. Soleus m.
5. Fibularis (peroneus) longus m.
6. Tendo calcaneus
7. Fibularis (peroneus) brevis m.
8. Medial malleolus of tibia
9. Lateral malleolus of fibula
10. Calcaneus
11. Abductor digiti minimi m.
12. Plantar surface of foot

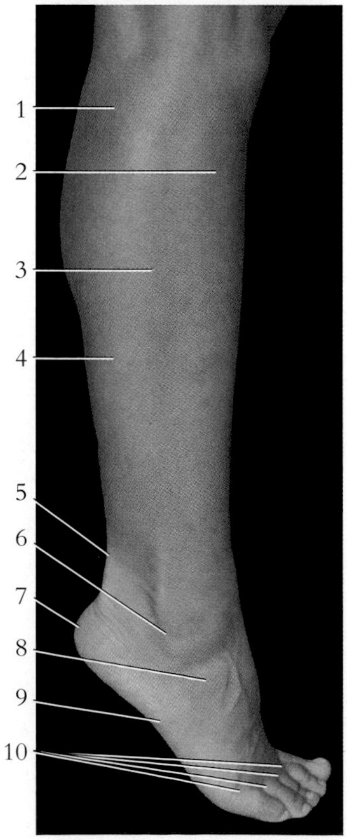

Figure 1.33 A lateral view of the right leg and foot.
1. Lateral head of gastrocnemius m.
2. Tibialis anterior m.
3. Peroneus longus m.
4. Soleus m.
5. Tendo calcaneus
6. Lateral malleolus of fibula
7. Calcaneus
8. Extensor digitorum brevis m.
9. Lateral surface of foot
10. Tendons of extensor digitorum longus m.

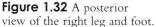

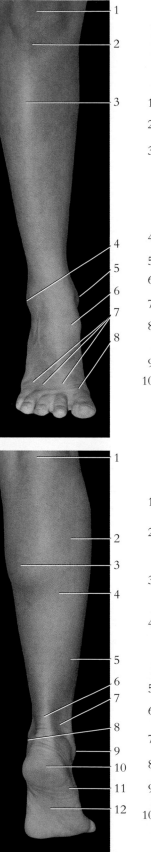

Cells

Cells are the basic structural and functional units of organization within the body. Although diverse, human cells have structural similarities including a **nucleus** containing a nucleolus, various **organelles** suspended in **cytoplasm**, and an enclosing cell (plasma) **membrane** (Fig. 2.1).

The **nucleus** is the large spheroid body within a cell that contains the **nucleoplasm**, one or more **nucleoli** and **chromatin**—the genetic material of the cell. The nucleus is enclosed by a double membrane called the **nuclear membrane**, or **nuclear envelope**. The **nucleolus** is a dense, nonmembranous body composed of protein and RNA molecules. The chromatin consists of proteins called histones and DNA molecules. Prior to cellular division, the chromatin shortens and coils into rod-shaped **chromosomes**.

The **cytoplasm** (or cytosol) of a cell is the medium of support between the nuclear membrane and the cell membrane. **Organelles** are minute structures within the cytoplasm of a cell that are concerned with specific functions. The cellular functions carried out by the organelles are referred to as **cellular metabolism**. The principal organelles and their functions are listed in Table 2.1. In order for cells to remain alive, metabolize, and maintain homeostasis, certain requirements must be met. These include having access to nutrients and oxygen, being able to eliminate wastes, and being maintained in a constant, protective environment.

The **cell membrane** is composed of phospholipid and protein molecules, which gives form to a cell and controls the passage of material into and out of a cell. More specifically, the proteins in the cell membrane provide: 1) structural support; 2) a mechanism of molecule transport across the membrane; 3) enzymatic control of chemical reactions; 4) receptors for hormones and other regulatory molecules; and 5) cellular markers (antigens), which identify the blood and tissue type. The carbohydrate molecules: 1) repel negative objects due to their negative charge; 2) act as receptors for hormones and other regulatory molecules; 3) form specific cell markers which enable like cells to attach and aggregate into tissues; and 4) enter into immune reactions.

The permeability of the cell membrane is a function of: 1) size of molecules; 2) solubility in lipids; 3) ionic charge of molecules; and 4) the presence of carrier molecules.

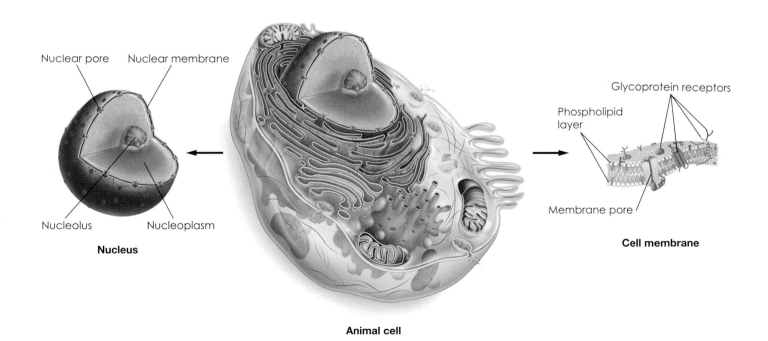

Nucleus

Nuclear pore Nuclear membrane

Nucleolus Nucleoplasm

Animal cell

Glycoprotein receptors

Phospholipid layer

Membrane pore

Cell membrane

Figure 2.1 A cell and its nucleus and cell (plasma) membrane.

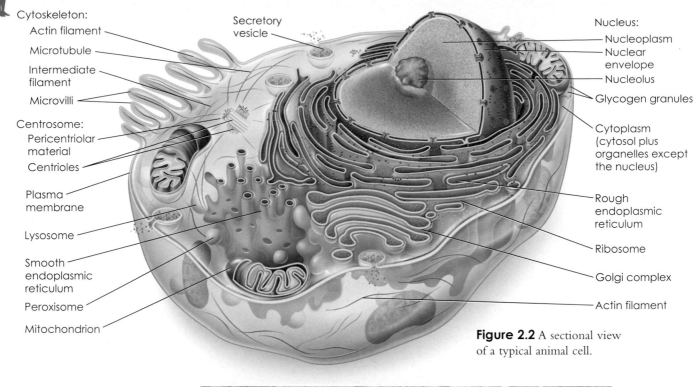

Cytoskeleton:
- Actin filament
- Microtubule
- Intermediate filament
- Microvilli

Centrosome:
- Pericentriolar material
- Centrioles

Plasma membrane

Lysosome

Smooth endoplasmic reticulum

Peroxisome

Mitochondrion

Secretory vesicle

Nucleus:
- Nucleoplasm
- Nuclear envelope
- Nucleolus

Glycogen granules

Cytoplasm (cytosol plus organelles except the nucleus)

Rough endoplasmic reticulum

Ribosome

Golgi complex

Actin filament

Figure 2.2 A sectional view of a typical animal cell.

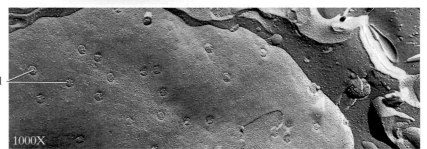

Figure 2.3 An electron micrograph of a freeze fractured nuclear envelope showing the nuclear pores.
1. Nuclear pores

1000X

Table 2.1 Structure and function of cellular components.

Component	Structure	Function
Cell (plasma) membrane	Composed of protein and phospholipid molecules	Provides form to cell and controls passage of materials into and out of cell
Cytoplasm	Fluid to jellylike substance	Suspends organelles; a matrix in which chemical reactions occur
Endoplasmic reticulum	Interconnecting hollow membranous channels	Facilitates cell transport of proteins; smooth ER produces lipids, rough ER produces proteins
Ribosomes	Granules of ribonucleic acid (RNA)	Synthesize proteins; located on rough endoplasmic reticulum
Mitochondria	Double-layered sacs with cristae	Production of ATP in aerobic respiration
Golgi complex	Stacked flattened sacs	Modifies, sorts, packages, and transports proteins from rough ER; forms secretory membrane and transport vesicles
Lysosomes	Membrane-surrounded sacs of enzymes	Digest foreign molecules and worn cells
Centrosome	Mass of two rodlike centrioles	Organizes spindle fibers and assists mitosis
Vessicles	Membranous sacs	Store and excrete substances within the cytoplasm
Cytoskeleton	Protein strands	Support cytoplasm and transport materials
Cilia and flagella	Cytoplasmic extensions from cell; contains microtubules	Movement of particles along cell surface or move cell
Nucleus	Nuclear membrane, nucleolus, and chromatin (DNA)	Directs cell activity; forms ribosomes

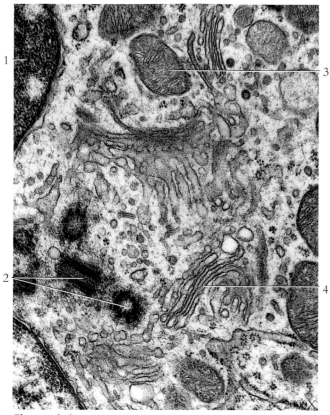

Figure 2.4 An electron micrograph of various organelles.

1. Nucleus
2. Centrioles
3. Mitochondrion
4. Golgi complex

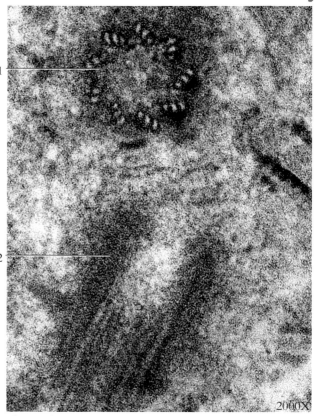

Figure 2.5 An electron micrograph of centrioles. The centrioles are positioned at right angles to one another.

1. Centriole (shown in cross section)
2. Centriole (shown in longitudinal section)

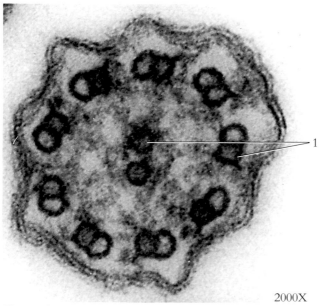

Figure 2.6 An electron micrograph of a cilium (cross section) showing the characteristic "9 + 2" arrangement of microtubules in the cross sections.

1. Microtubules

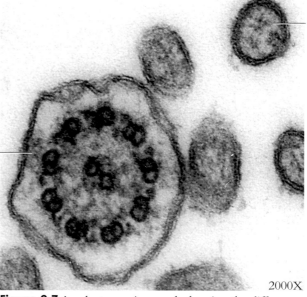

Figure 2.7 An electron micrograph showing the difference between a microvillus and a cilium.

1. Cilium
2. Microvillus

Electron micrographs courtesy of Scott C. Miller

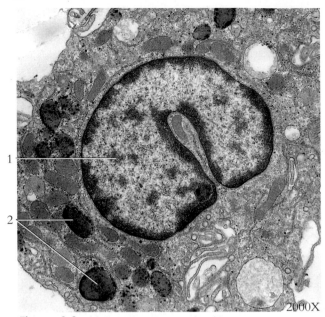

Figure 2.8 An electron micrograph of lysosomes.
1. Nucleus
2. Lysosomes

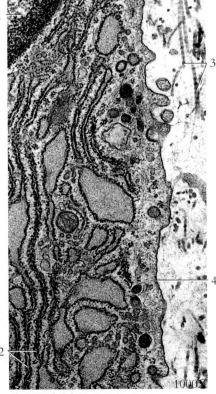

Figure 2.9 An electron micrograph of a mitochondrion.
1. Outer membrane   3. Inner membranes
2. Crista

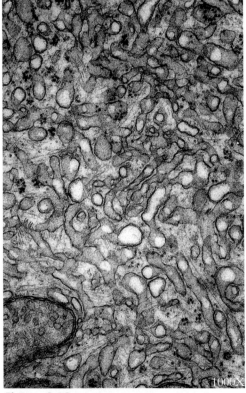

Figure 2.10 An electron micrograph of smooth endoplasmic reticulum from the testis.

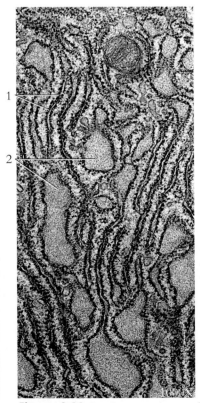

Figure 2.11 An electron micrograph of rough endoplasmic reticulum.
1. Ribosomes
2. Cisternae

Figure 2.12 Rough endoplasmic reticulum secreting collagenous filaments to the outside of the cell.
1. Nucleus
2. Rough endoplasmic reticulum
3. Collagenous filaments
4. Cell membrane

Electron micrographs courtesy of Scott C. Miller

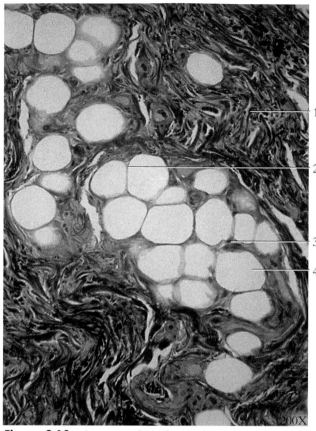

Figure 2.13 Adipocytes (fat cells) in adipose tissue.
1. Loose connective tissue
2. Cell membrane of adipocyte
3. Nucleus
4. Lipid-filled vacuole of adipocyte

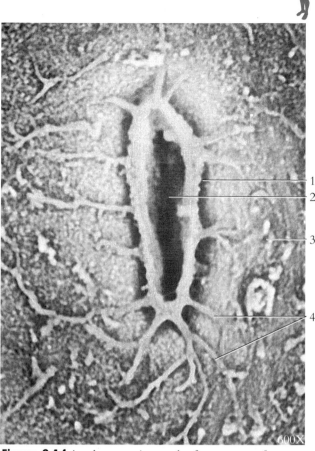

Figure 2.14 An electron micrograph of an osteocyte (bone cell) in cortical bone matrix.
1. Lacuna   3. Bone matrix
2. Osteocyte   4. Canaliculi

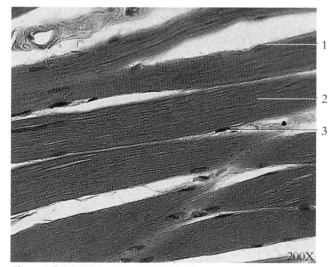

Figure 2.15 Skeletal muscle cells (fibers).
1. Sarcolemma (cell   2. Striations
   membrane)   3. Nucleus

Figure 2.16 An electron micrograph of an erythrocyte (red blood cell).

Electron micrographs courtesy of Scott C. Miller

15

Figure 2.17 An electron micrograph of a skeletal muscle myofibril, showing the striations.

1. Mitochondria
2. Z line
3. A band
4. I band
5. T tubule
6. Sarcoplasmic reticulum
7. H band
8. Sacromere

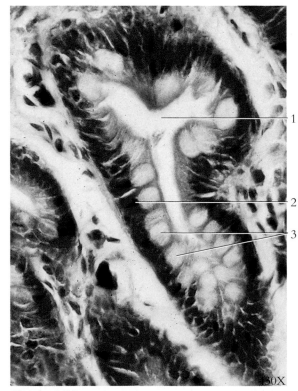

Figure 2.18 Goblet cells within an intestinal gland (crypt of Lieberkühn) of small intestine.

1. Lumen of gland
2. Nucleus
3. Goblet cells

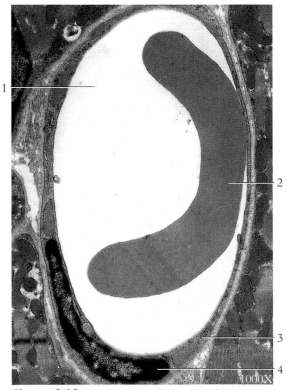

Figure 2.19 An electron micrograph of a capillary containing an erythrocyte.

1. Lumen of capillary
2. Erythrocyte
3. Endothelial cell
4. Nucleus of endothelial cell

Electron micrographs courtesy of Scott C. Miller

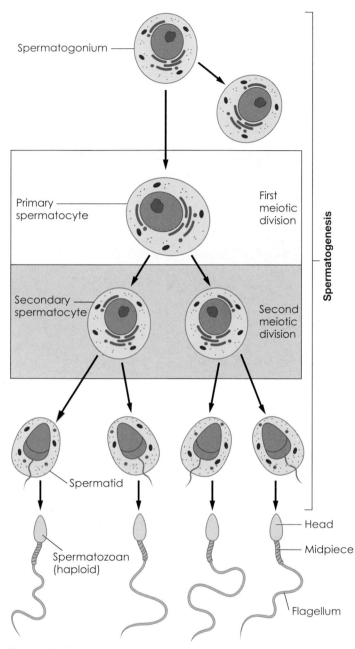

Spermatogonium

Primary spermatocyte — First meiotic division

Secondary spermatocyte — Second meiotic division

Spermatogenesis

Spermatid

Spermatozoan (haploid)

Head

Midpiece

Flagellum

Figure 2.20 Spermatogenesis is the production of male gametes, or spermatozoa, through the process of meiosis.

Oogonium

Primary oocyte — **Meiosis I**

Polar body

Secondary oocyte

Sperm contacts secondary oocyte

Polar bodies degenerate

Meiosis II

Fertilization of female gamete (ova) with male gamete

Maturation

Zygote (diploid)

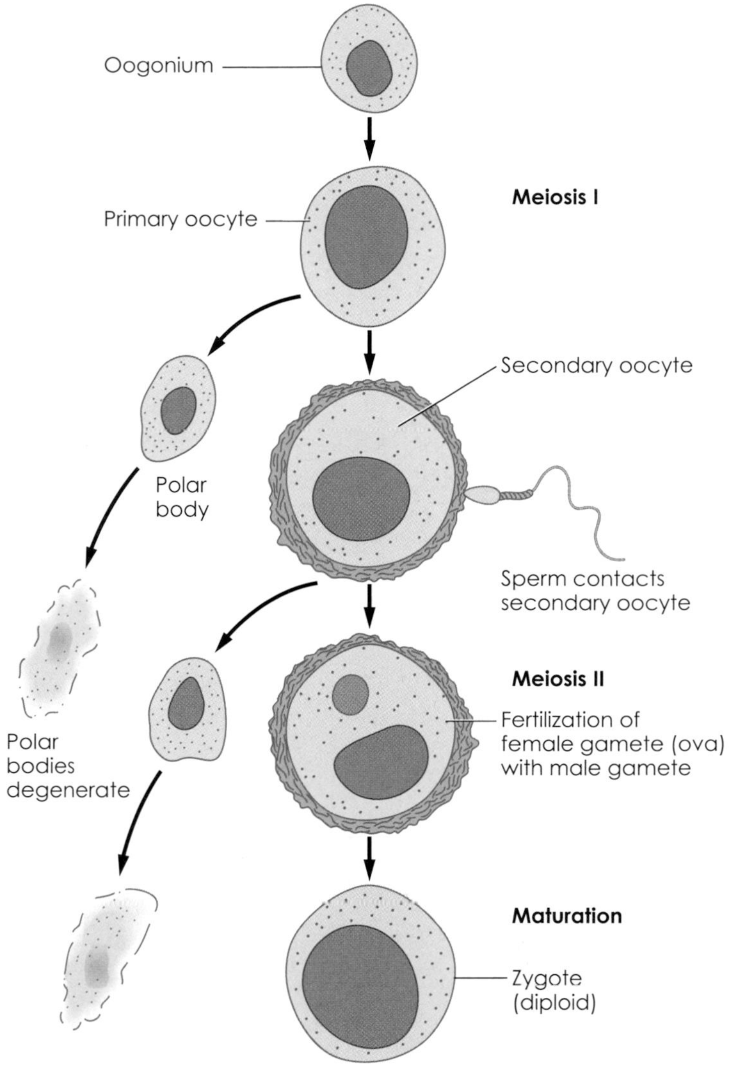

Figure 2.21 Oogenesis is the production of female gametes, or ova, through the process of meiosis.

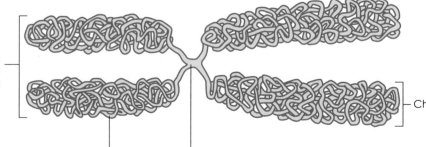

One (duplicated) chromosome composed of two identical chromatids

Chromatid

Chromatin strand   Centromere

Figure 2.22 Each duplicated chromosome consists of two identical chromatids attached at the centrally located and constricted centromere.

Figure 2.23 Stages of mitosis.   (All photographs on this page are 800X.)

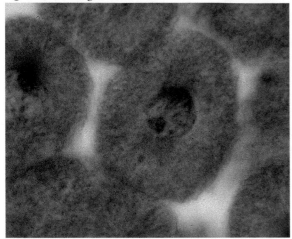

Prophase Each chromosome consists of two chromatids joined by a centromere. Spindle fibers extend from each centriole.

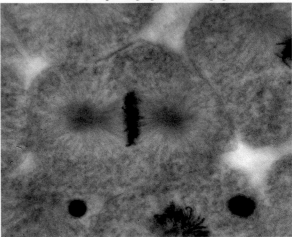

Metaphase The chromosomes are positioned at the equator. The spindle fibers from each centriole are attached to the centromeres.

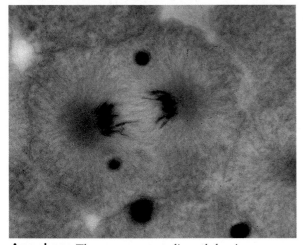

Anaphase The centromeres split, and the sister chromatids separate as each is pulled to an opposite pole.

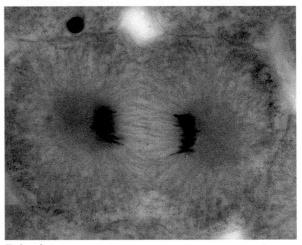

Telophase The chromosomes begin to uncoil and become less distinct. The cell membrane forms between the forming daughter cells.

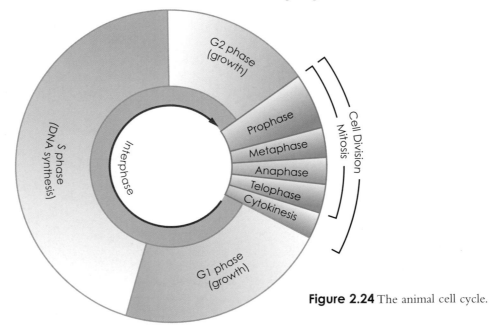

G2 phase (growth)

S phase (DNA synthesis)

Interphase

Prophase

Metaphase

Anaphase

Telophase

Cytokinesis

Cell Division Mitosis

G1 phase (growth)

Figure 2.24 The animal cell cycle.

Histology

While it is true that cells comprise the basic structural and functional units of the body, the cells in a multicellular organism, such as a human, are so specialized that they do not function independently. **Tissues** are aggregations of similar cells that perform specific functions. **Histology** is the science concerned with the study of tissues. Both **cytology**, the study of cells, and histology are actually microscopic anatomy. Although cytologists and histologists utilize many different techniques to study cells and tissues, basically only two kinds of microscopes are used to view the prepared specimens. **Light microscopy** is used for the general observation of cellular and tissue structure, and **electron microscopy** permits observation of the fine details of the specimens.

In electron microscopy, a beam of electrons is passed through an object in a procedure called **transmission electron microscopy (TEM)**, or the beam is reflected off the surface of an object in a procedure called **scanning electron microscopy (SEM)**. In both cases, the electron beam is magnified with electromagnets. The depth of focus of SEM is much greater than it is with TEM, producing a clear three-dimensional image of cellular or tissue structure. The magnification ability of SEM, however, is not as great as that of TEM.

The tissues of the body are classified into four principal types, determined by structure and function: 1) **epithelial tissues** cover body and organ surfaces, line body and luminal (hollow portion of body tube) cavities, and form various glands; 2) **connective tissues** bind, support, and protect body parts; 3) **muscle tissues** contract to produce movements; and 4) **nervous tissues** initiate and transmit nerve impulses from one body part to another.

Epithelial tissues are classified by the number of layers of cells and the shapes of the cells along the exposed (apical) surfaces. A **simple epithelial tissue** is made up of a single layer of cells. A **stratified epithelial tissue** is made up of layers of cells. The basic shapes of the exposed cells are: **squamous**, or flattened; **cuboidal**, or cube-shaped; and **columnar**, or elongated.

Connective tissues are classified according to the characteristics of the **matrix**, or binding material between the similar cells. The classification of connective tissues is not exact, but the following is a commonly accepted scheme of classification:

A. **Embryonic connective tissue**
B. **Connective tissue proper**
    1. Loose (areolar) connective tissue
    2. Dense regular connective tissue
    3. Dense irregular connective tissue
    4. Elastic connective tissue
    5. Reticular connective tissue
    6. Adipose tissue
C. **Cartilage**
    1. Hyaline cartilage
    2. Fibrocartilage
    3. Elastic cartilage
D. **Bone tissue**
E. **Blood (vascular tissue)**

Muscle tissues are responsible for the movement of materials through the body, the movement of one part of the body with respect to another, and for locomotion. The three kinds of muscle tissue are **smooth**, **cardiac**, and **skeletal**. The fibers in all three kinds are adapted to contract in response to stimuli.

Nervous tissues are composed of **neurons**, which respond to stimuli and conduct action potentials (nerve impulses) to and from all body organs, and **neuroglia**, which functionally support and physically bind neurons.

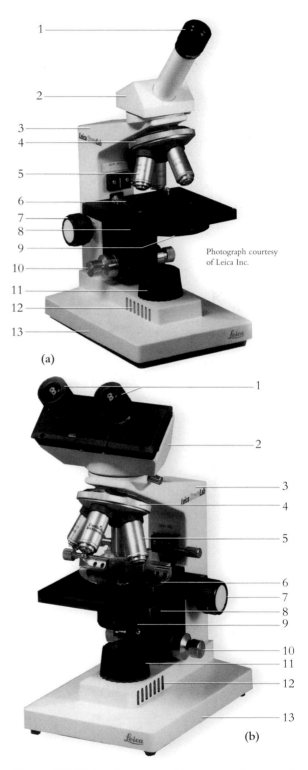

Photograph courtesy of Leica Inc.

(a)

(b)

Figure 3.1 Light microscopes, (a) compound monocular microscope, and (b) compound binocular microscope.

1. Eyepiece (ocular)
2. Body
3. Arm
4. Nosepiece
5. Objective
6. Stage clip
7. Focus adjustment knob
8. Stage
9. Condenser
10. Fine focus adjustment knob
11. Collector lens with field diaphragm
12. Illuminator (inside)
13. Base

Epithelial Tissue

Epithelial tissue covers the outside of the body and lines all organs. Its primary function is to provide protection.

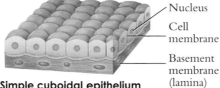
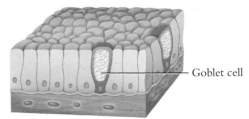

Nucleus
Cell membrane
Basement membrane (lamina)

Goblet cell

Simple squamous epithelium   Simple cuboidal epithelium   Simple columnar epithelium

Figure 3.2 Examples of animal epithelial tissues.

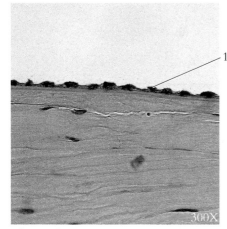

Figure 3.3 Simple squamous epithelium.
1. Single layer of flattened cells

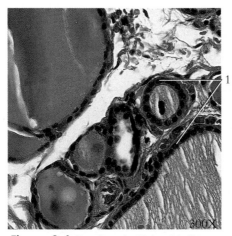

Figure 3.4 Simple cuboidal epithelium.
1. Single layer of cells with round nuclei

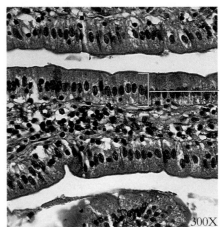

Figure 3.5 Simple columnar epithelium.
1. Single layer of cells with oval nuclei

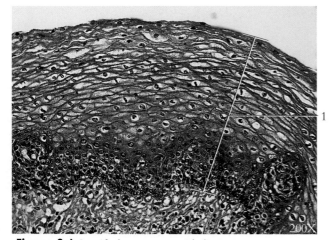

Figure 3.6 Stratified squamous epithelium.
1. Multiple layers of cells, which are flattened at the upper layer

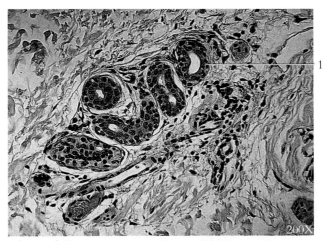

Figure 3.7 Stratified cuboidal epithelium.
1. Two layers of cells with round nuclei

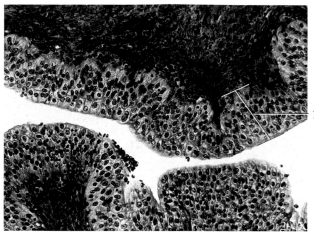

Figure 3.8 Transitional epithelium.
1. Cells are balloon-like at surface

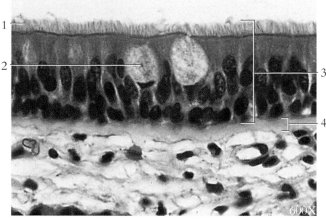

Figure 3.9 Pseudostratified columnar epithelium.
1. Cilia
2. Goblet cell
3. Pseudostratified columnar epithelium
4. Basement membrane

Connective Tissue

Connective tissue functions as a binding and supportive tissue for all other tissues in the organism.

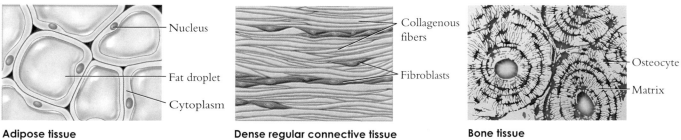

Adipose tissue — Nucleus, Fat droplet, Cytoplasm

Dense regular connective tissue — Collagenous fibers, Fibroblasts

Bone tissue — Osteocyte, Matrix

Figure 3.10 Examples of animal connective tissues.

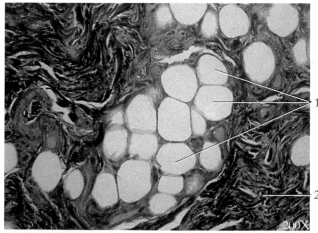

Figure 3.11 Adipose connective tissue.
1. Adipocytes (adipose cells)
2. Loose connective tissue

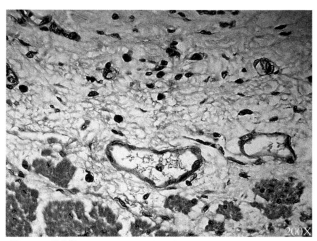

Figure 3.12 Loose connective tissue.

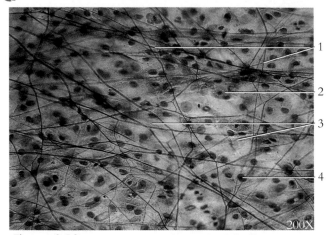

Figure 3.13 Loose connective tissue stained for fibers.
1. Elastic fibers (black)   3. Collagen fibers (pink)
2. Mast cell   4. Fibroblast

Figure 3.14 Dense regular connective tissue.
1. Nuclei of fibroblasts arranged in parallel rows

Electron micrographs courtesy of Scott C. Miller

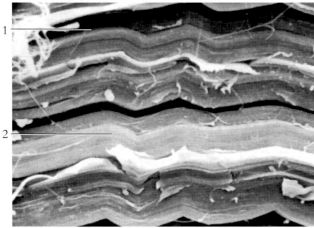

Figure 3.15 An electron micrograph of dense regular connective tissue.
1. Collagenous fiber
2. Fibroblast

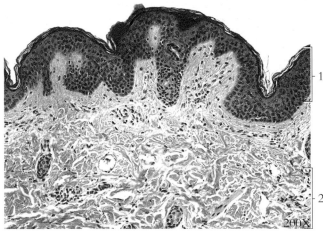

Figure 3.16 Dense irregular connective tissue.
1. Epidermis
2. Dense irregular connective tissue (reticular layer of dermis)

Figure 3.17 An electron micrograph of dense irregular connective tissue.
1. Collagenous fibers

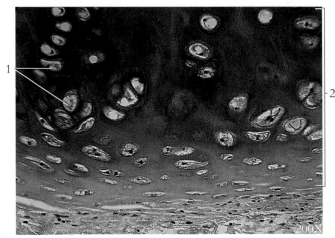

Figure 3.18 Hyaline cartilage.
1. Chondrocytes
2. Hyaline cartilage

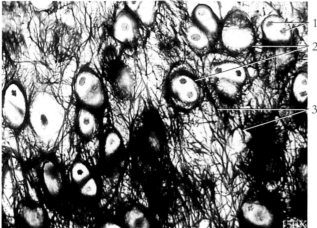

Figure 3.19 Elastic cartilage.
1. Chondrocytes   3. Elastic fibers
2. Lacunae

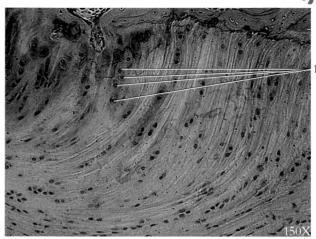

Figure 3.20 Fibrocartilage.
1. Chondrocytes arranged in a row

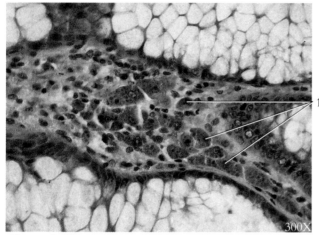

Figure 3.21 Cells of connective tissue.
1. Macrophages

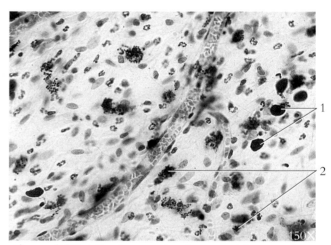

Figure 3.22 Cells of connective tissue, special preparation.
1. Mast cells   2. Macrophages

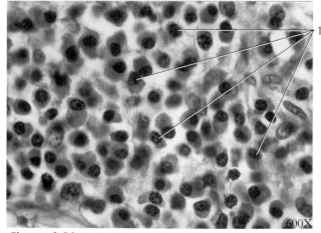

Figure 3.23 Cells of connective tissue.
1. Plasma cells

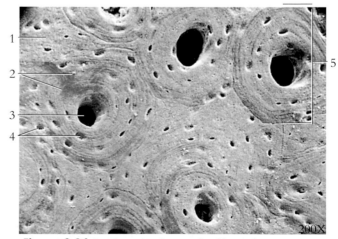

Figure 3.24 An electron micrograph of bone tissue.
1. Interstitial lamellae   4. Lacunae
2. Lamellae   5. Osteon (haversian
3. Central canal (haversian canal)   system)

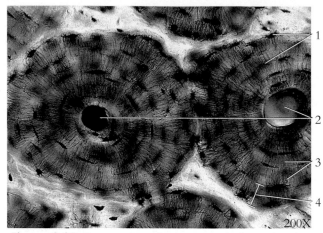

Figure 3.25 Cross section of two osteons in bone tissue.
1. Osteocytes within lacunae
2. Central (haversian) canals
3. Canaliculi
4. Lamella

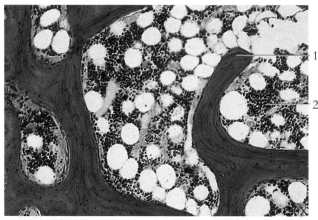

Figure 3.26 Spongy (cancellous) bone.
1. Spicule of bone
2. Red marrow (includes adipose cells)

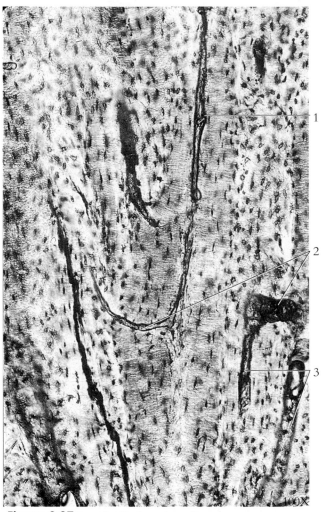

Figure 3.27 Longitudinal section of osteons.
1. Central (haversian) canal
2. Perforating (Volkman's) canals
3. Central (haversian) canals

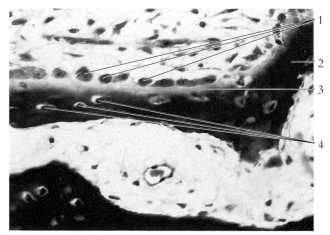

Figure 3.28 Osteoblasts.
1. Osteoblasts
2. Bone
3. Osteoid
4. Osteocytes

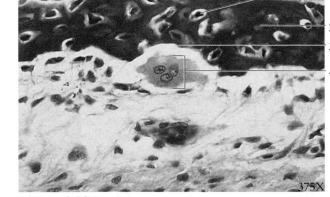

Figure 3.29 Osteoclast.
1. Osteocytes
2. Bone
3. Howship's lacuna
4. Osteoclast in Howship's lacuna

Muscle Tissue

Muscle tissue is a tissue adapted to contract. Muscles provide movement and functionality to the organism.

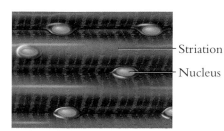

— Striation

— Nucleus

Skeletal muscle

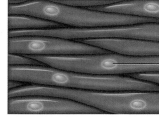

— Intercalated disc

— Striation

— Nucleus

Cardiac muscle

— Nucleus

Smooth muscle

Figure 3.30 Examples of animal muscle tissues.

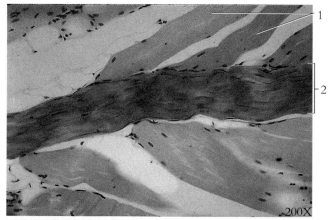

Figure 3.31 Longitudinal section of skeletal muscle tissue.
1. Skeletal muscle cells, note striations
2. Multiple nuclei in periphery of cell

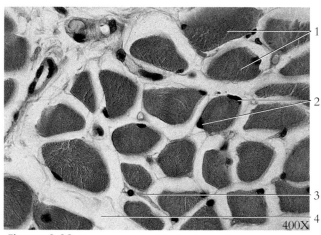

Figure 3.32 Cross section of skeletal muscle tissue.
1. Skeletal muscle cells
2. Nuclei in periphery of cell
3. Endomysium (surrounds cells)
4. Perimysium (surrounds bundles of cells)

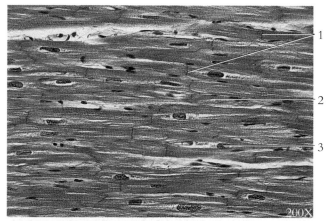

Figure 3.33 Attachment of skeletal muscle to tendon.
1. Skeletal muscle
2. Dense regular connective tissue (tendon)

Figure 3.34 Cardiac muscle tissue.
1. Intercalated discs
2. Light-staining perinuclear sarcoplasm
3. Nucleus in center of cell

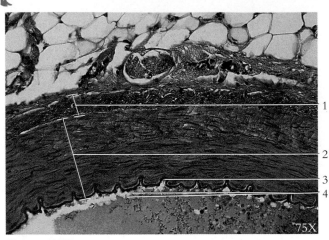

Figure 3.35 Wall of muscular artery.
1. Tunica adventitia
2. Tunica media
3. Internal elastic membrane
4. Endothelial cells

Figure 3.36 Partially teased smooth muscle tissue.
1. Nucleus of individual cell

Nervous Tissue

Nervous tissue functions to receive stimuli and transmits signals from one part of the organism to another.

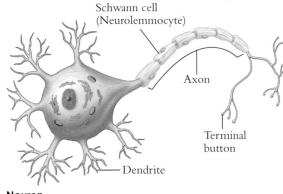

Schwann cell
(Neurolemmocyte)

Axon

Terminal
button

Dendrite

Neuron

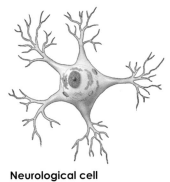

Neurological cell

Figure 3.37 Examples of animal nervous tissues.

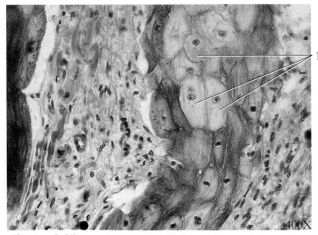

Figure 3.38 Conduction myofibers (Purkinje fibers).
1. Conduction myofibers in the heart

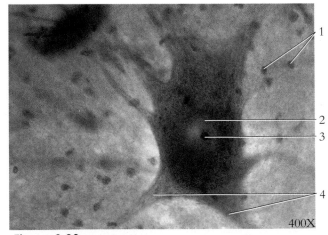

Figure 3.39 Nervous tissue.
1. Nuclei of surrounding neuroglial cells
2. Nucleus of neuron
3. Nucleolus of neuron
4. Dendrites of neuron

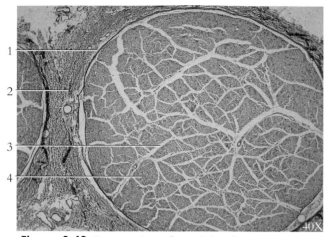

Figure 3.40 Cross section of a peripheral nerve.
1. Perineurium   3. Endoneurium
2. Epineurium   4. Bundle of axons

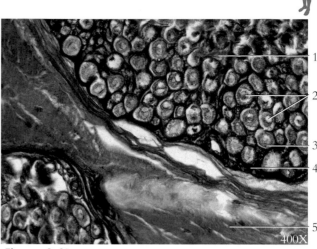

Figure 3.41 Cross section of a peripheral nerve.
1. Endoneurium   3. Myelin sheath   5. Epineurium
2. Axons   4. Perineurium

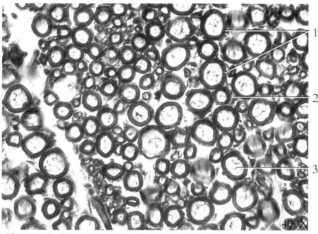

Figure 3.42 A nerve stained with osmium.
1. Myelin sheath
2. Endoneurium
3. Axon

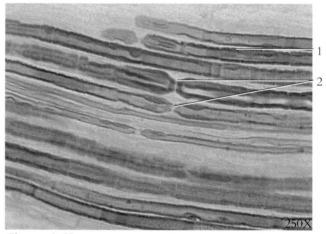

Figure 3.43 A longitudinal section of axons.
1. Myelin sheath
2. Neurofibril nodes (nodes of Ranvier)

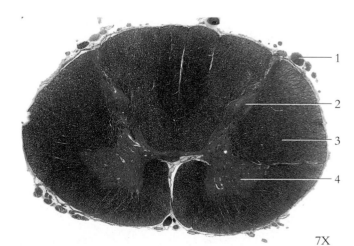

Figure 3.44 Cross section of the spinal cord.
1. Posterior (dorsal) root of spinal nerve
2. Posterior (dorsal) horn (gray matter)
3. Spinal cord tract (white matter)
4. Anterior (ventral) horn (gray matter)

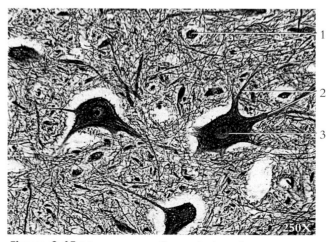

Figure 3.45 Motor neurons from spinal cord.
1. Neuroglia cells
2. Dendrites
3. Nucleus

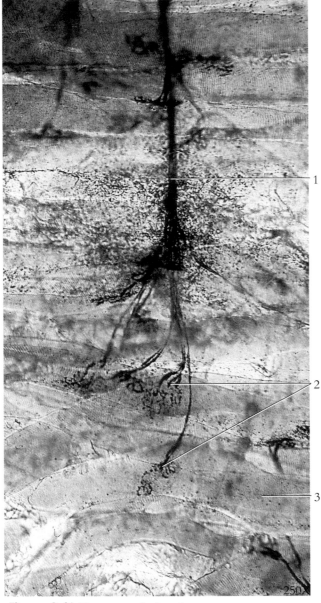

Figure 3.46 Neuromuscular junction.
1. Motor nerve   3. Skeletal muscle fiber
2. Motor end plates

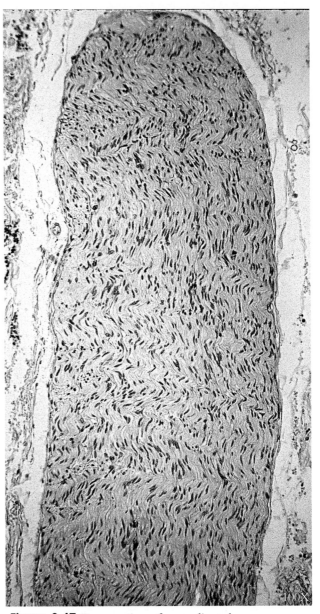

Figure 3.47 Cross section of unmyelinated nerve.

Integumentary System

Chapter 4

The integumentary system consists of the **integument**, or **skin**, and its associated **hair**, **glands**, and **nails** (Fig. 4.1). The skin is composed of an outer **epidermis** consisting of numerous layers and a **dermis** consisting of two layers. The **hypodermis (subcutaneous tissue)** connects the skin to the underlying organs.

The keratinized squamous epithelium of the epidermis is stratified into layers. From superficial to deep, they are the **stratum corneum**, the **stratum lucidum** (only in skin of the palms and soles), the **stratum granulosum**, the **stratum spinosum**, and the **stratum basale (stratum germinativum)**. The strata basale undergos mitosis (cell division). Pigments, such as **melanin**, are found in the stratum basale and the protein **keratin** is found in all but the deepest epidermal layers. Both are protective. The stratum corneum is **cornified** (hardened and scale-like) for further protection.

The dermis is divisible into the **stratum papillarosum** (papillary layer) and the **stratum reticularosum** (reticular layer). The hypodermis is the deep, binding layer of adipose and loose connective tissue.

The skin provides several important functions. The skin protects the body from disease and external injury. Keratin and an acidic, oily secretion on the surface protect the skin from water and microorganisms. Cornification protects against abrasion, and **melanin** (a dark pigment) is a barrier to UV light. The skin regulates body fluids and temperatures by radiation, convection, and sweating. It permits the absorption of some UV light, respiratory gases, steroids, and fat-soluble vitamins. The skin synthesizes melanin and keratin, which remain in the skin, and begins synthesis of vitamin D, which is used elsewhere in the body. The skin provides sensory reception through cutaneous receptors and provides development and growth of hair and certain exocrine glands.

Formed prenatally as invaginations of the epidermis into the dermis, hair, glands, and nails provide protection to the skin. Each hair develops in a **hair follicle** and is protective against sunlight and mild abrasions. Integumentary glands are classified as **sebaceous** (oil secreting), **sudoriferous** (sweat), and **ceruminous** (wax-producing). (**Mammary glands** are specialized sweat glands that produce milk in a lactating female.) A nail protects the terminal end of each digit. The fingernails also aid in picking up objects and scratching.

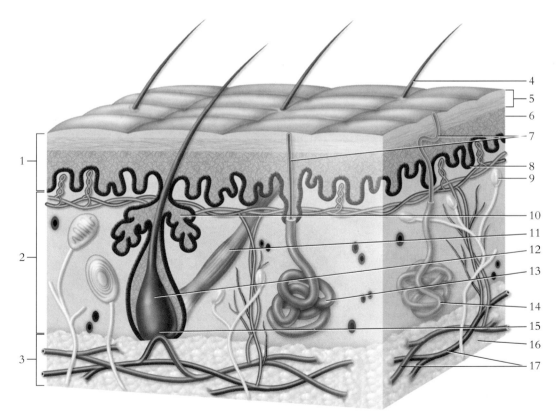

Figure 4.1 The skin and certain epidermal structures.

1. Epidermis
2. Dermis
3. Hypodermis
4. Shaft of hair
5. Stratum corneum and granulosum
6. Stratum spinosum
7. Sweat duct
8. Stratum basale
9. Sensory receptor
10. Sebaceous gland
11. Arrector pili muscle
12. Hair follicle
13. Apocrine sweat gland
14. Eccrine sweat gland
15. Bulb of hair
16. Adipose tissue
17. Cutaneous blood vessels

Figure 4.2 The gross structure of the skin and underlying fascia.
1. Epidermis   4. Fascia
2. Dermis   5. Muscle
3. Hypodermis

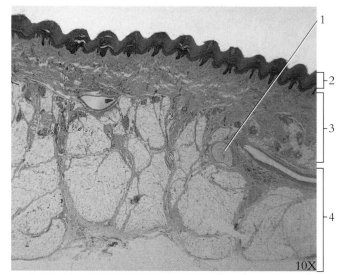

Figure 4.3 Skin.
1. Lamellated (Pacinian) corpuscle   3. Dermis
2. Epidermis   4. Hypodermis

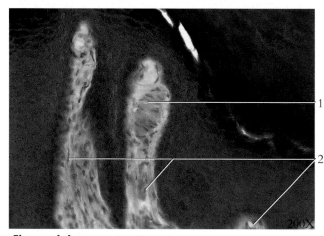

Figure 4.4 Corpuscle of touch.
1. Corpuscle of touch (Meissner's corpuscle)
2. Dermal papillae

Figure 4.5 Thick skin.
1. Stratum corneum   4. Stratum spinosum
2. Stratum lucidum   5. Stratum basale
3. Stratum granulosum   6. Dermis

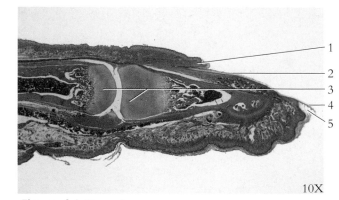

Figure 4.6 Fingertip.
1. Eponychium   4. Free border of nail
2. Nail plate   5. Hyponychium
3. Phalanges

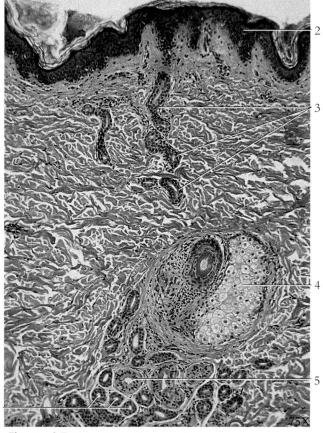

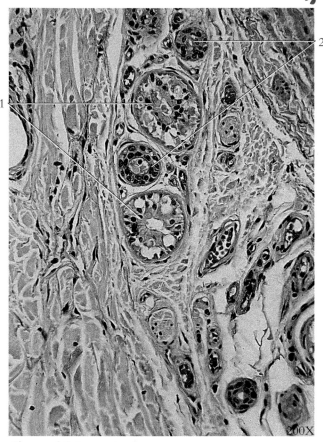

Figure 4.7 Thin skin with sweat gland.
1. Excretory duct portion of sweat gland
2. Epidermis
3. Excretory duct of sweat gland (coiling toward surface)
4. Sebaceous gland
5. Secretory portion of sweat gland

Figure 4.8 Sweat gland.
1. Secretory portion
   (large diameter with light–staining columnar cells)
2. Excretory portion
   (small diameter with dark–staining stratified cuboidal cells)

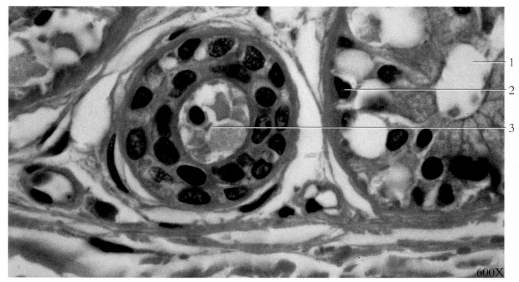

Figure 4.9 Sweat gland.
1. Lumen of secretory portion
2. Myoepithelial cell
3. Lumen of excretory portion (or duct)

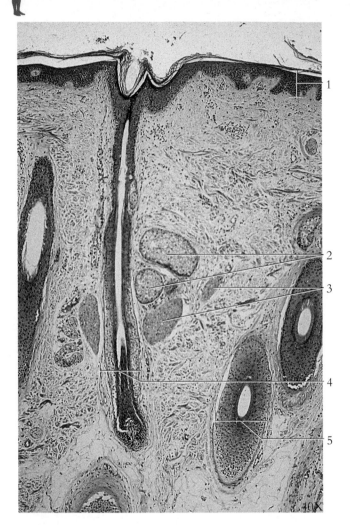

Figure 4.10 Hair follicle.
1. Epidermis
2. Sebaceous glands
3. Arrector pili muscle
4. Hair follicle
5. Hair follicle (oblique cut)

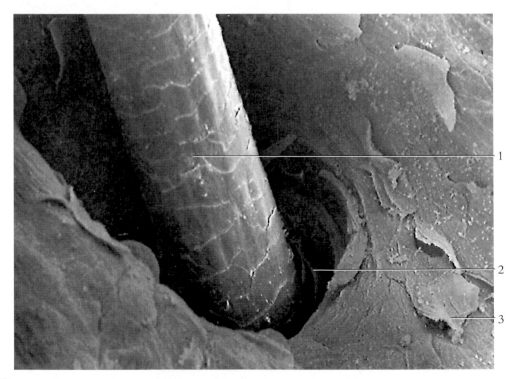

Figure 4.11 An electron micrograph of a hair emerging from a hair follicle.
1. Shaft of hair
   (note the scalelike pattern)
2. Hair follicle
3. Epithelial cell from
   stratum corneum

Skeletal System: Axial

Chapter 5

The skeletal system of an adult human is composed of approximately 206 bones—the number varies from person to person depending on genetic variations. Some adults have extra bones in the skull called **sutural (wormian) bones**. Additional bones may develop in tendons as the tendons move across a joint. Bones formed this way are called **sesamoid bones**, and the patella (kneecap) is an example.

The skeleton is divided into axial and appendicular portions (Table 5.1). The **axial skeleton** consists of the bones that form the axis of the body and that support and protect the organs of the head, neck, and trunk. The axial skeleton includes the bones of the skull, auditory ossicles, hyoid bone, vertebral column, and rib cage.

The **appendicular skeleton** (see Chapter 6) is composed of the bones of the upper and lower extremities and the bony girdles, which anchor the appendages to the axial skeleton. The appendicular skeleton includes the bones of the pectoral girdle, upper extremities, pelvic girdle, and lower extremities.

The mechanical functions of the bones of the skeleton include the support and protection of softer body tissues and organs. Also, certain bones function as levers during body movement. The metabolic functions of bones include **hematopoiesis**, or manufacture of blood cells and platelets, and mineral storage. Calcium and phosphorus are the two principal minerals stored within bone and give bone its rigidity and strength.

The bones of the skeleton are classified into four principal types on the basis of shape rather than size. The four classes of bones are **long bones, short bones, flat bones,** and **irregular bones** (Fig. 5.1).

Table 5.1 Classifications of the bones of the adult skeleton.

Axial Skeleton

Rib cage 25 bones	24 ribs 1 sternum
Vertebral column 26 bones	7 cervical vertebra 12 thoracic vertebra 5 lumbar vertebra 1 sacrum (5 fused bones) 1 coccyx (3–5 fused bones)
Skull 22 bones	
14 facial bones	2 maxilla 2 palatine bones 2 zygomatic bones 2 lacrimal bones 2 nasal bones 1 vomer 2 inferior nasal concha 1 mandible
8 cranial bones	1 frontal bone 2 parietal bones 1 occipital bone 2 temporal bones 1 sphenoid bone 1 ethmoid bone
Auditory ossicles 6 bones	2 malleus 2 incus 2 stapes
Hyoid 1 bone	1 hyoid

Appendicular Skeleton

Pectoral girdle 4 bones	2 scapulae 2 clavicles
Pelvic girdle 2 bones	2 os coxae (each contains 3 fused bones: ilium, ischium, and pubis)
Upper extremities 60 bones	2 humerus 2 radius 2 ulnas 16 carpal bones 10 metacarpal bones 28 phalanx
Lower extremities 60 bones	2 femurs 2 patellas 2 tibia 2 fibulas 14 tarsal bones 10 metatarsal bones 28 phalanx

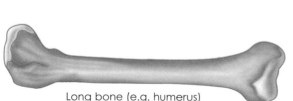

Long bone (e.g. humerus)

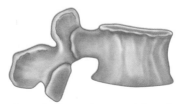

Short bone (e.g. carpals)

Irregular bone (e.g. vertebrae)

Flat bone (e.g. cranial bones)

Figure 5.1 The shapes of bones.

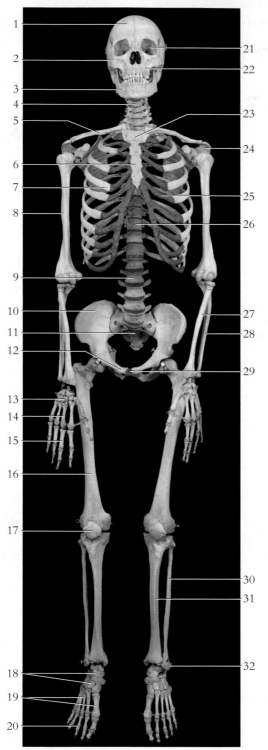

Figure 5.2 An anterior view of the skeleton.

1. Frontal bone
2. Zygomatic bone
3. Mandible
4. Cervical vertebra
5. Clavicle
6. Body of sternum
7. Rib
8. Humerus
9. Lumbar vertebra
10. Ilium
11. Sacrum
12. Pubis
13. Carpal bones
14. Metacarpal bones
15. Phalanges
16. Femur
17. Patella
18. Tarsal bones
19. Metatarsal bones
20. Phalanges
21. Orbit
22. Maxilla
23. Manubrium
24. Scapula
25. Costal cartilage
26. Thoracic vertebra
27. Radius
28. Ulna
29. Symphysis pubis
30. Fibula
31. Tibia
32. Lateral malleolus

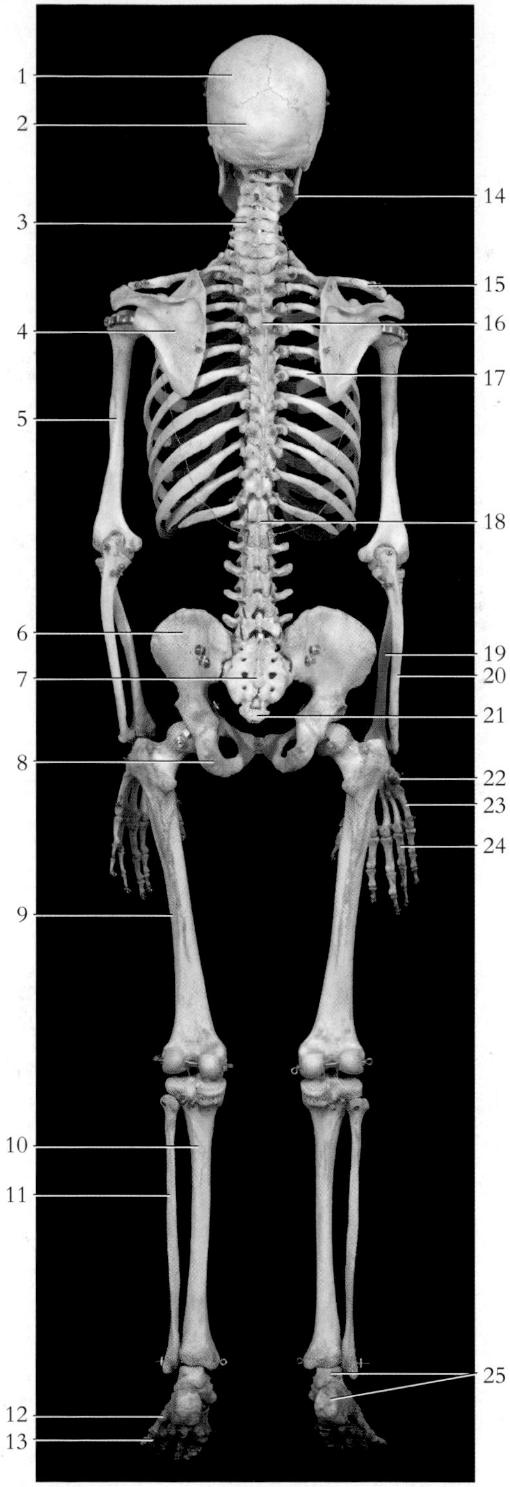

Figure 5.3 A posterior view of the skeleton.

1. Parietal bone
2. Occipital bone
3. Cervical vertebra
4. Scapula
5. Humerus
6. Ilium
7. Sacrum
8. Ischium
9. Femur
10. Tibia
11. Fibula
12. Metatarsal bones
13. Phalanges
14. Mandible
15. Clavicle
16. Thoracic vertebra
17. Rib
18. Lumbar vertebra
19. Radius
20. Ulna
21. Coccyx
22. Carpal bones
23. Metacarpal bones
24. Phalanges
25. Tarsal bones

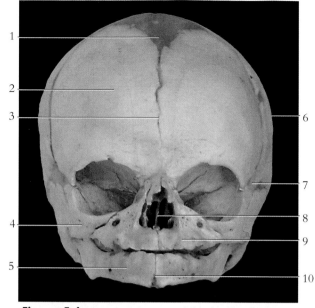

Figure 5.4 An anterior view of the fetal skull.
1. Anterior fontanel
2. Frontal bone
3. Frontal suture
4. Zygomatic bone
5. Mandible
6. Parietal bone
7. Anterolateral fontanel
8. Nasal septum
9. Maxilla
10. Mental symphysis

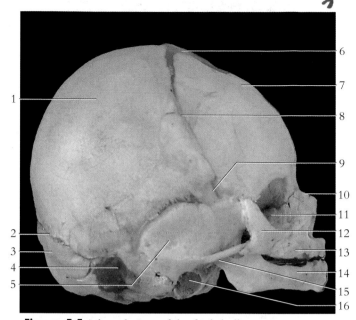

Figure 5.5 A lateral view of the fetal skull.
1. Parietal bone
2. Area of lambdoidal suture
3. Occipital bone
4. Posterolateral fontanel
5. Temporal bone
6. Anterior fontanel
7. Frontal bone
8. Area of coronal suture
9. Anterolateral fontanel
10. Nasal bone
11. Ethmoid bone
12. Zygomatic bone
13. Maxilla
14. Mandible
15. Zygomatic process
16. External acoustic meatus

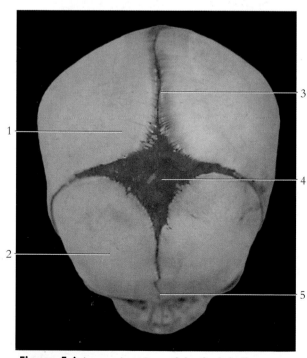

Figure 5.6 A superior view of the fetal skull.
1. Parietal bone
2. Frontal bone
3. Area of sagittal suture
4. Anterior fontanel
5. Frontal suture

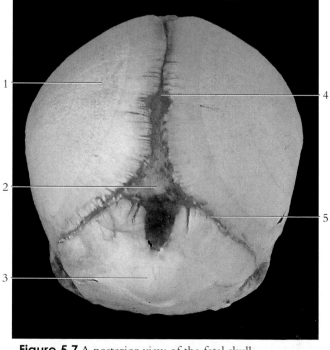

Figure 5.7 A posterior view of the fetal skull.
1. Parietal bone
2. Posterior fontanel
3. Occipital bone
4. Area of sagittal suture
5. Area of lambdoidal suture

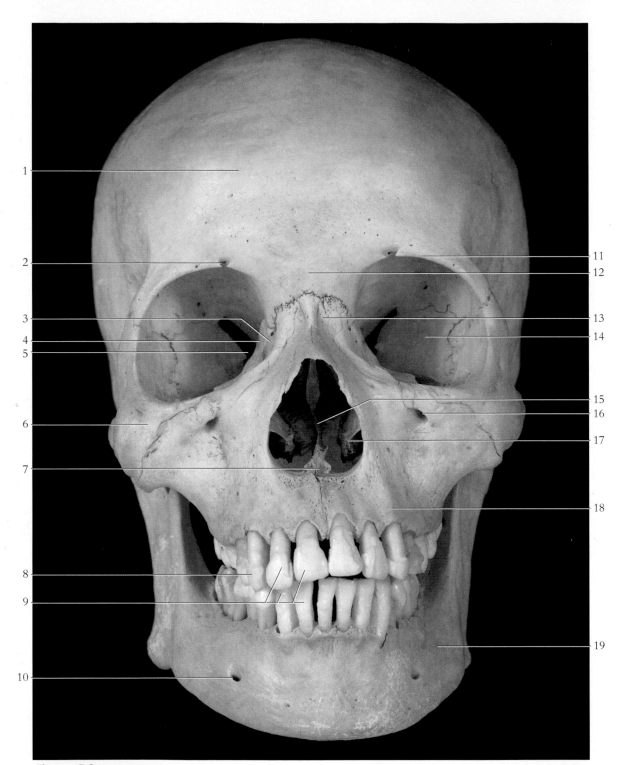

Figure 5.8 An anterior view of the skull.

1. Frontal bone
2. Supraorbital foramen
3. Lacrimal bone
4. Orbital plate of ethmoid bone
5. Superior orbital fissure
6. Zygomatic bone
7. Vomer

8. Canine
9. Incisors
10. Mental foramen
11. Supraorbital margin
12. Glabella
13. Nasal bone
14. Sphenoid bone

15. Perpendicular plate of ethmoid bone
16. Infraorbital foramen
17. Inferior nasal concha
18. Maxilla
19. Mandible

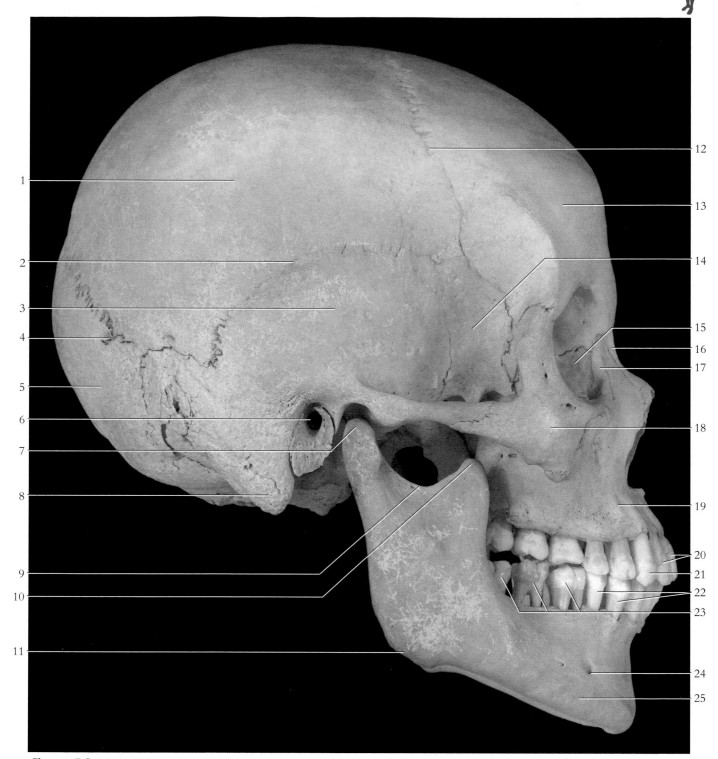

Figure 5.9 A lateral view of the skull.

1. Parietal bone
2. Squamosal suture
3. Temporal bone
4. Lambdoidal suture
5. Occipital bone
6. External acoustic meatus
7. Mandibular condyle
8. Mastoid process of temporal bone
9. Mandibular notch
10. Coronoid process of mandible
11. Angle of mandible
12. Coronal suture
13. Frontal bone
14. Sphenoid bone
15. Orbital plate of ethmoid bone
16. Nasal bone
17. Lacrimal bone
18. Zygomatic bone
19. Maxilla
20. Incisors
21. Canine
22. Premolars
23. Molars
24. Mental foramen
25. Mandible

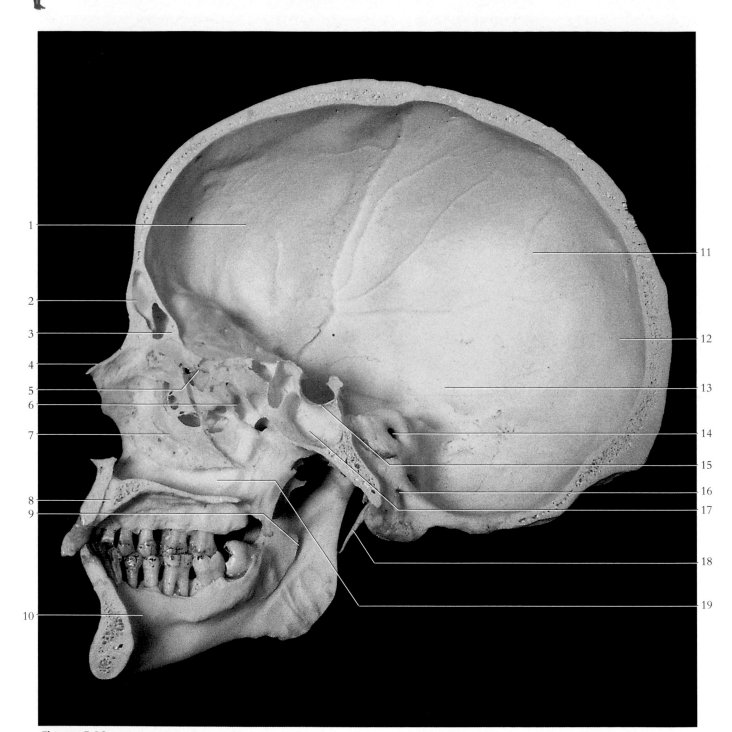

Figure 5.10 A sagittal view of the skull.

1. Frontal bone
2. Frontal sinus
3. Crista galli of ethmoid bone
4. Nasal bone
5. Cribriform plate of ethmoid bone
6. Ethmoidal sinus
7. Nasal concha (inferior)
8. Maxilla
9. Mandibular foramen
10. Mandible
11. Parietal bone
12. Occipital bone
13. Temporal bone
14. Internal acoustic meatus
15. Sella turcica
16. Hypoglossal canal
17. Sphenoidal sinus
18. Styloid process of temporal bone
19. Vomer

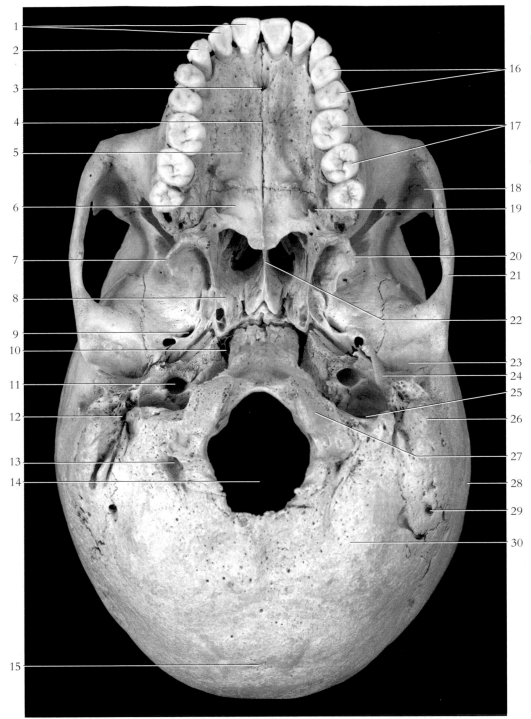

Figure 5.11 An inferior view of the skull.

1. Incisors
2. Canine
3. Incisive foramen
4. Median palatine suture
5. Maxilla (palatine process)
6. Palatine bone (horizontal plate)
7. Lateral pterygoid process of sphenoid bone
8. Sphenoid bone
9. Foramen ovale
10. Foramen lacerum
11. Carotid canal
12. Stylomastoid foramen
13. Condyloid canal
14. Foramen magnum
15. External occipital protuberance
16. Premolars
17. Molars
18. Zygomatic bone
19. Greater palatine foramen
20. Lateral pterygoid process of sphenoid bone
21. Zygomatic arch
22. Vomer
23. Mandibular fossa
24. Styloid process of temporal bone
25. Jugular foramen
26. Mastoid process of temporal bone
27. Occipital condyle
28. Temporal bone
29. Mastoid foramen
30. Occipital bone

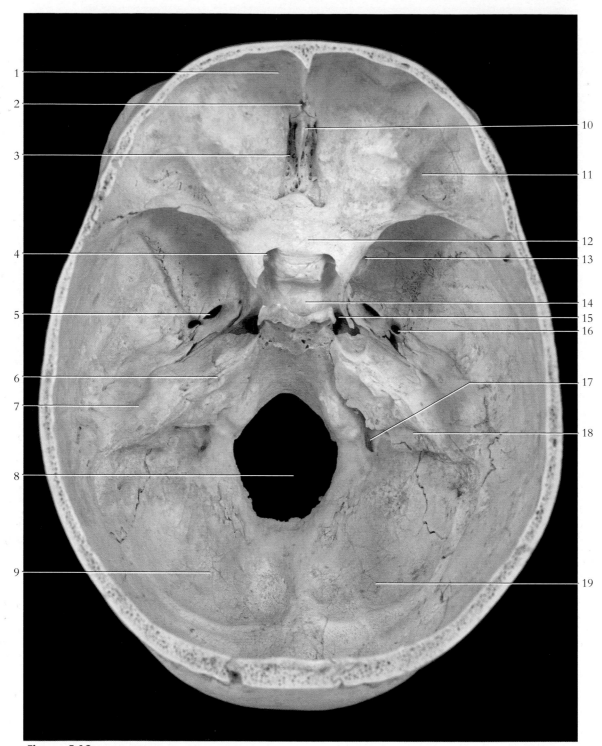

Figure 5.12 A superior view of the cranium.

1. Frontal bone
2. Foramen cecum
3. Cribriform plate with olfactory foramina
4. Optic canal
5. Foramen ovale
6. Petrous part of temporal bone
7. Temporal bone
8. Foramen magnum
9. Occipital bone
10. Crista galli of ethmoid bone
11. Anterior cranial fossa
12. Sphenoid bone
13. Foramen rotundum
14. Sella turcica of sphenoid bone
15. Foramen lacerum
16. Foramen spinosum
17. Jugular foramen
18. Internal acoustic meatus
19. Posterior cranial fossa

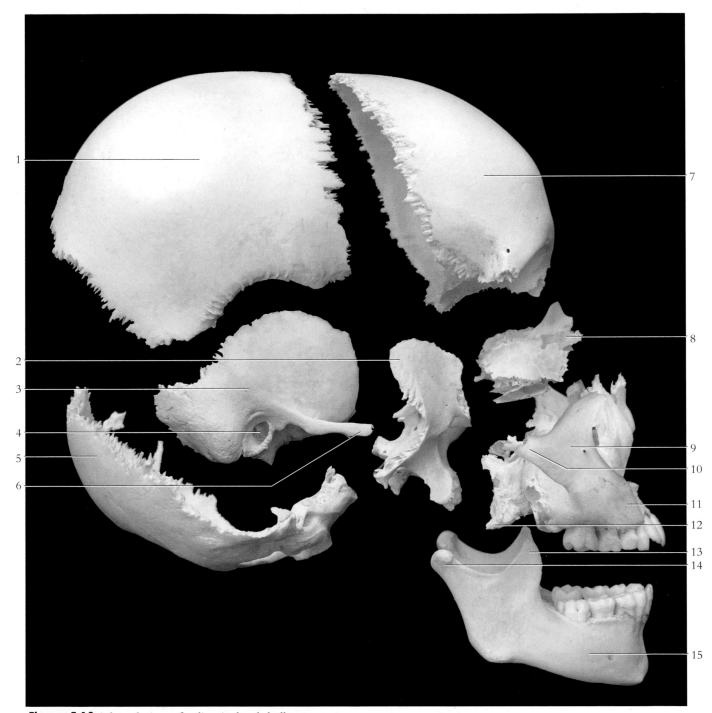

Figure 5.13 A lateral view of a disarticulated skull.

1. Parietal bone
2. Sphenoid bone
3. Temporal bone
4. External acoustic meatus
5. Occipital bone
6. Zygomatic process
   of temporal bone
7. Frontal bone
8. Ethmoid bone
9. Zygomatic bone
10. Temporal process of
    zygomatic bone
11. Maxilla
12. Palatine bone
13. Coronoid process
    of mandible
14. Condylar process
    of mandible
15. Mandible

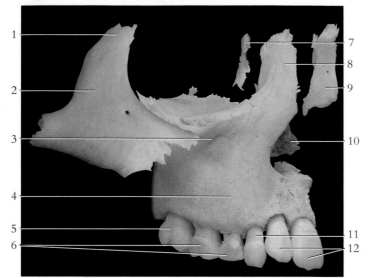

Figure 5.14 The bones of the right facial region.
1. Orbital process of zygomatic bone
2. Zygomatic bone
3. Infraorbital foramen
4. Maxilla
5. Molar
6. Premolars
7. Lacrimal bone
8. Frontal process of maxilla
9. Nasal bone
10. Inferior nasal concha
11. Canine
12. Incisors

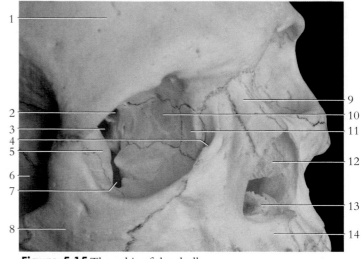

Figure 5.15 The orbit of the skull.
1. Frontal bone
2. Optic foramen
3. Superior orbital fissure
4. Lacrimal foramen
5. Sphenoid bone
6. Temporal bone
7. Inferior orbital fissure
8. Zygomatic bone
9. Nasal bone
10. Ethmoid bone
11. Lacrimal bone
12. Inferior nasal concha
13. Vomer
14. Maxilla

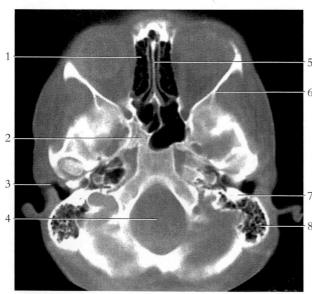

Figure 5.16 A CT transaxial section through the skull.
1. Nasal cavity
2. Sphenoid bone
3. External acoustic meatus
4. Foramen magnum
5. Nasal septum
6. Wall of bony orbit
7. Middle ear chamber
8. Mastoidal sinus

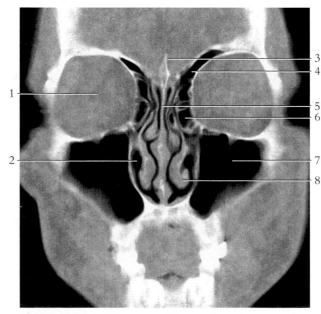

Figure 5.17 A CT image of the nasal cavity and paranasal sinuses.
1. Eyeball in orbit
2. Nasal cavity
3. Crista galli
4. Frontal sinus
5. Perpendicular plate of ethmoid bone
6. Ethmoidal sinus
7. Maxillary sinus
8. Inferior nasal concha

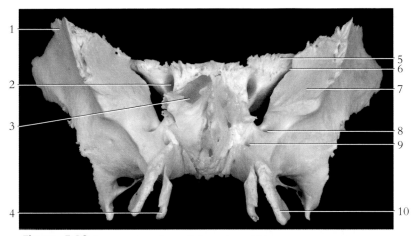

Figure 5.18 An anterior view of the sphenoid bone.

1. Greater wing of sphenoid bone
2. Optic foramen
3. Opening into sphenoidal sinus
4. Medial pterygoid plate
5. Lesser wing of sphenoid bone
6. Superior orbital fissure
7. Orbital surface of greater wing of sphenoid bone
8. Foramen rotundum
9. Pterygoid canal
10. Lateral pterygoid plate

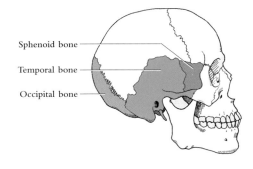

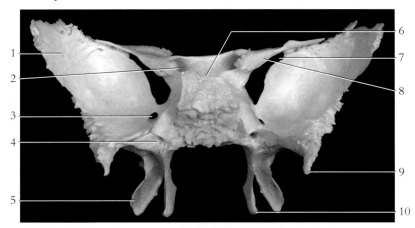

Figure 5.19 A posterior view of the sphenoid bone.

1. Greater wing of sphenoid bone
2. Optic foramen
3. Foramen rotundum
4. Pterygoid canal
5. Lateral pterygoid plate
6. Superior orbital fissure
7. Sella turcica
8. Lesser wing of sphenoid bone
9. Spine of sphenoid bone
10. Medial pterygoid plate

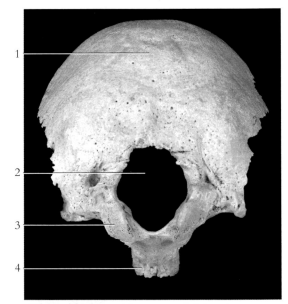

Figure 5.20 An inferior view of the occipital bone.

1. External occipital protuberance
2. Foramen magnum
3. Occipital condyle
4. Pharyngeal tubercle

Figure 5.21 A lateral view of the right temporal bone.

1. External acoustic meatus
2. Mastoid part of temporal bone
3. Mastoid process
4. Squamous part of temporal bone
5. Zygomatic process of temporal bone
6. Mandibular fossa
7. Tympanic part of temporal bone

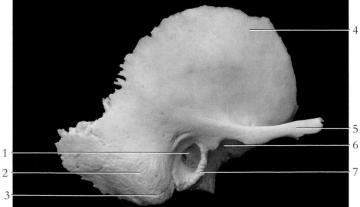

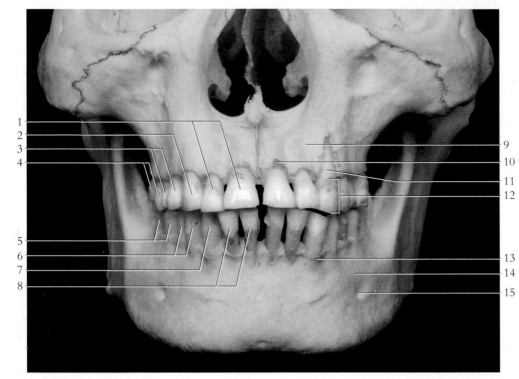

Figure 5.22 An anterior view of the jaws and teeth.
1. Superior incisors
2. Superior canine
3. Superior premolars
4. Superior molars
5. Inferior molars
6. Inferior premolars
7. Inferior canine
8. Inferior incisors
9. Maxilla
10. Dental alveolus
11. Root of tooth
12. Crown of tooth
13. Alveolar process
14. Mandible
15. Mental foramen

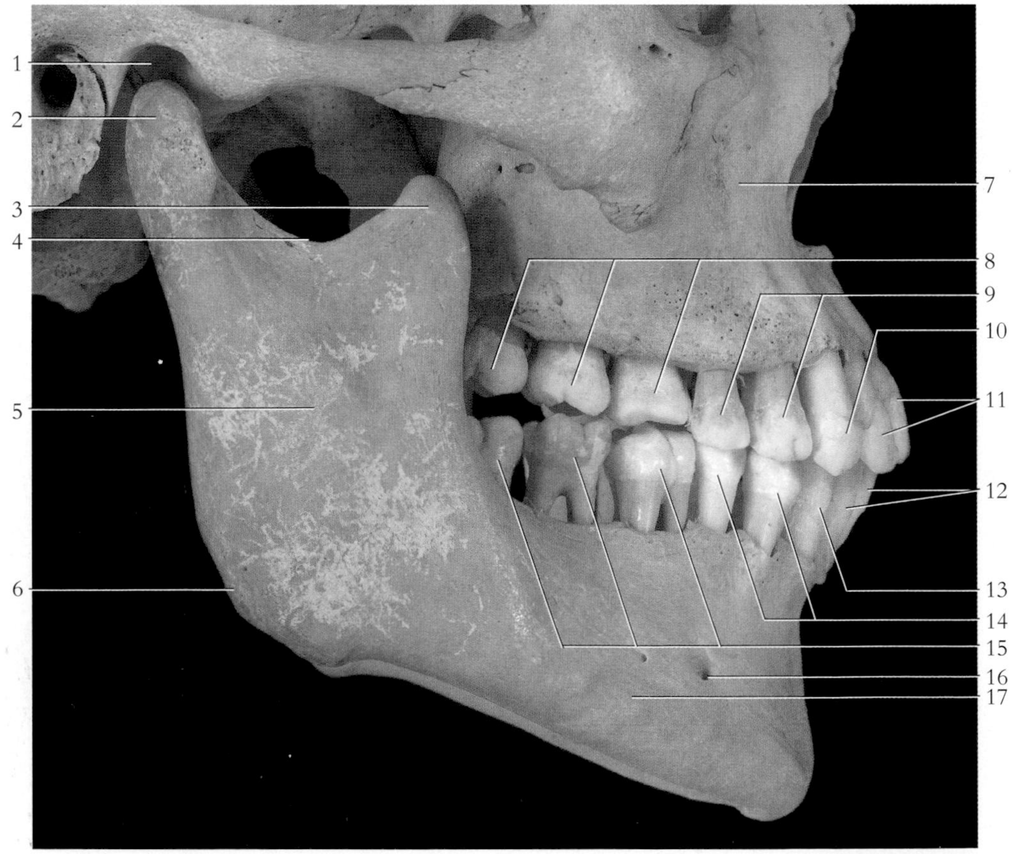

Figure 5.23 A lateral view of the jaws and teeth.
1. Mandibular fossa (half of temporomandibular joint)
2. Mandibular condyl (other half of temporomandibular joint)
3. Coronoid process of mandible
4. Mandibular notch
5. Ramus of mandible
6. Angle of mandible
7. Maxilla
8. Superior molars
9. Superior premolars
10. Superior canine
11. Superior incisors
12. Inferior incisors
13. Inferior canine
14. Inferior premolars
15. Inferior molars
16. Mental foramen
17. Body of mandible

Figure 5.24 The eruption of teeth seen in a dissected skull of a youth (9 to 12 years old).

1. Permanent second molar
2. Permanent first molar
3. Deciduous second molar
4. Permanent first molar
5. Permanent second molar
6. Permanent second premolar
7. Permanent second premolar
8. Permanent canine
9. Deciduous first molar
10. Permanent first premolar
11. Deciduous canine
12. Incisors
13. Deciduous canine
14. Deciduous first molar
15. Permanent canine
16. Permanent first premolar

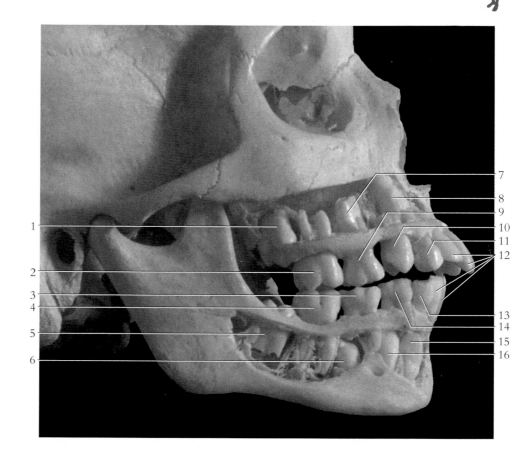

Figure 5.25 A medial view of the jaws and teeth.

1. Ethmoidal sinus
2. Inferior nasal concha
3. Inferior meatus
4. Palatine portion of maxilla
5. Maxilla
6. Superior molars
7. Superior premolars
8. Incisors
9. Inferior premolars
10. Inferior molars
    (note impacted wisdom tooth)
11. Mandible
12. Sphenoidal sinus
13. Styloid process of temporal bone
14. Medial pterygoid process

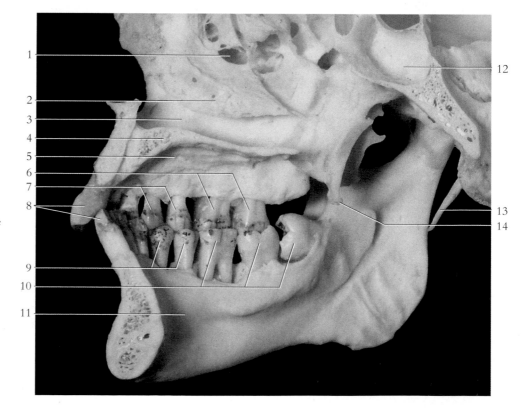

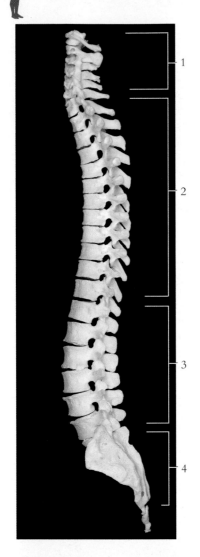

Figure 5.26 The curvature of the vertebral column.
1. Cervical curvature (7 vertebrae)
2. Thoracic curvature (12 vertebrae)
3. Lumbar curvature (5 vertebrae)
4. Pelvic (sacral) curvature (5 fussed vertebrae)

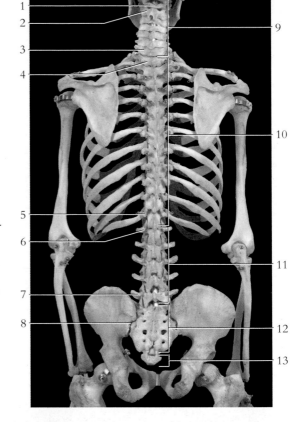

Figure 5.27 A posterior view of the vertebral column.
1. Atlas
2. Axis
3. Seventh cervical vertebra
4. First thoracic vertebra
5. Twelfth thoracic vertebra
6. First lumbar vertebra
7. Fifth lumbar vertebra
8. Sacroiliac joint
9. Cervical vertebrae
10. Thoracic vertebrae
11. Lumbar vertebrae
12. Sacrum
13. Coccyx

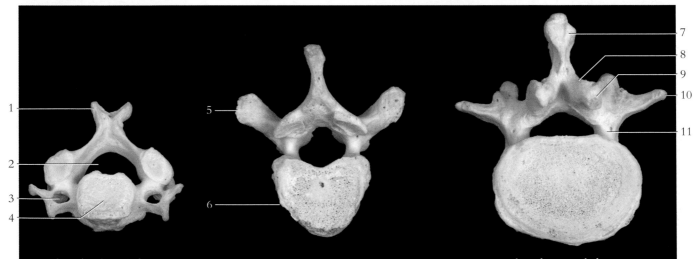

Cervical vertebra

Thoracic vertebra

Lumbar vertebra

Figure 5.28 Representative vertebrae (inferior view).
1. Spinous process (note that it is bifid)
2. Vertebral foramen
3. Transverse foramen (in transverse process)
4. Body of vertebrae
5. Transverse process
6. Inferior facet for head of rib
7. Spinous process
8. Lamina
9. Inferior articular surface
10. Transverse process
11. Pedicle

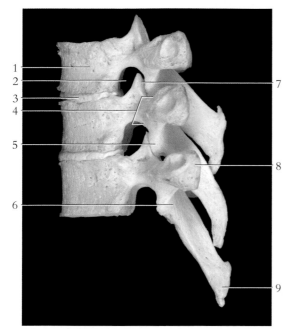

Figure 5.29 A lateral view of thoracic vertebrae.

1. Vertebral body
2. Intervertebral foramen
3. Intervertebral disc
4. Pedicle
5. Inferior articular process
6. Lamina
7. Superior articular facet
8. Transverse process
9. Spinous process

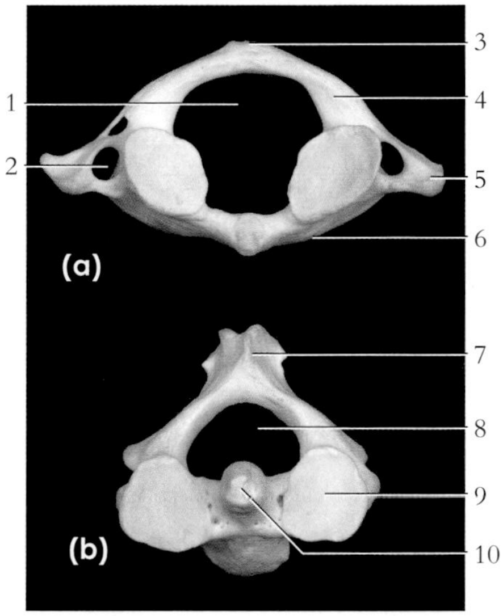

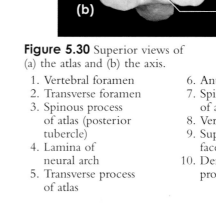

Figure 5.30 Superior views of (a) the atlas and (b) the axis.

1. Vertebral foramen
2. Transverse foramen
3. Spinous process of atlas (posterior tubercle)
4. Lamina of neural arch
5. Transverse process of atlas
6. Anterior arch of atlas
7. Spinous process of axis
8. Vertebral foramen
9. Superior articular facet for atlas
10. Dens (odontoid process)

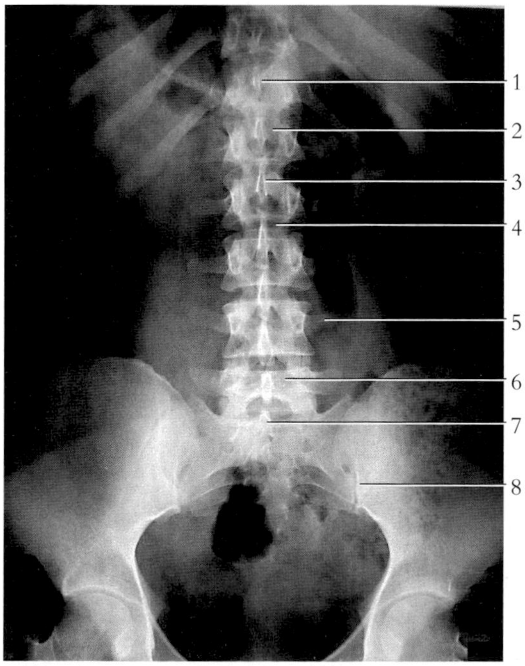

Figure 5.31 A radiograph of the lumbar vertebrae.

1. T12
2. Body of L1
3. Spinous process of L2
4. Intervertebral disc
5. Transverse process of L4
6. Lamina of L5
7. Sacrum
8. Sacroiliac joint

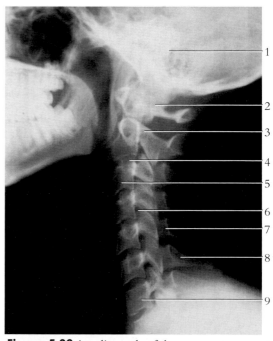

Figure 5.32 A radiograph of the cervical vertebrae.

1. Petrous portion of temporal bone
2. Atlas
3. Axis
4. Intervetebral disc
5. Body of C3
6. Intervertebral foramen
7. Spinous process of C5
8. Spinous process of C6
9. Body of C7

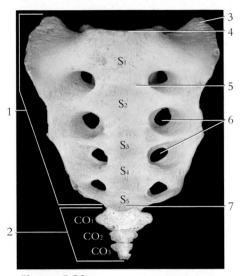

Figure 5.33 An anterior view of the sacrum and coccyx.
1. Sacrum
2. Coccyx
3. Sacral ala
4. Base of sacrum
5. Transverse line
6. Anterior sacral foramina
7. Apex of sacrum

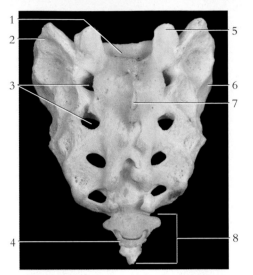

Figure 5.34 A posterior view of the sacrum and coccyx.
1. Superior portion of sacral canal
2. Sacral ala
3. Posterior sacral foramina
4. Coccygeal cornu
5. Superior articular process
6. Auricular surface
7. Median sacral crest
8. Coccyx

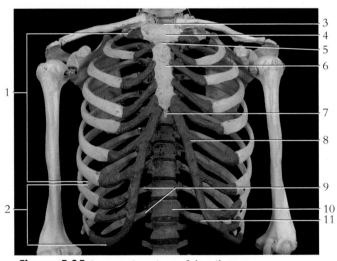

Figure 5.35 An anterior view of the rib cage.
1. True ribs (seven pairs)
2. False ribs (five pairs)
3. Jugular notch
4. Manubrium
5. Sternal angle
6. Body of sternum
7. Xiphoid process
8. Costal cartilage
9. Floating ribs (inferior two pairs of false ribs)
10. Twelfth thoracic vertebra
11. Twelfth rib

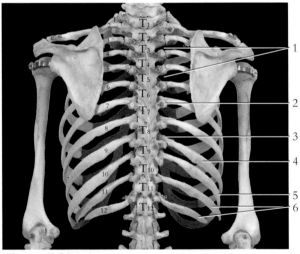

Figure 5.36 A posterior view of the rib cage.
1. Intercostal spaces
2. Transverse process of thoracic vertebra
3. Tubercle of rib
4. Angle of rib
5. Body of rib
6. Floating ribs

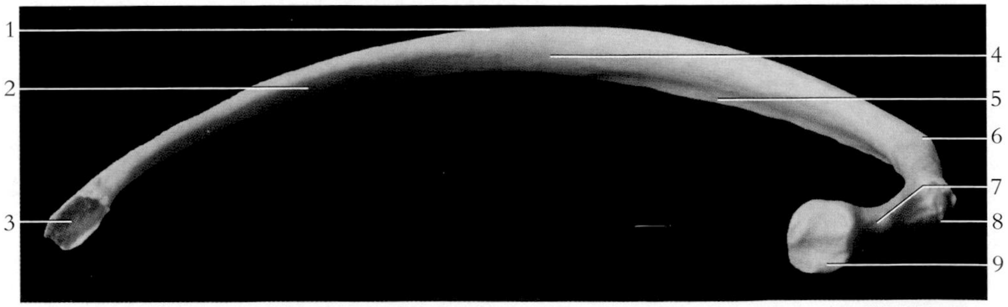

Figure 5.37 A rib.
1. Superior border
2. Body of rib
3. Articulation site for costal cartilage
4. Internal surface
5. Costal groove
6. Angle of rib
7. Neck of rib
8. Articular surface
9. Head of rib

Skeletal System: Appendicular

The structure of the **pectoral girdle** and **upper extremities** of the appendicular skeleton is adaptive for freedom of movement and extensive muscle attachment. The structure of the **pelvic girdle** and **lower extremities** is adaptive for support and locomotion.

The pectoral girdle is composed of two **scapulae** and two **clavicles**. The clavicles attach the pectoral girdle to the axial skeleton at the **sternum**. The bones of each upper extremity are the **humerus, ulna, radius**, eight **carpal bones**, five **metacarpal bones**, and fourteen **phalanges**. In addition, two or more **sesamoid bones** at the interphalangeal joints are usually present.

The pelvic girdle, or **pelvis**, is formed by two **os coxae** (hip bones) which are united anteriorly by the **symphysis pubis**. The pelvic girdle is attached posteriorly to the sacrum of the vertebral column. Each **os coxae** consists of three separate bones: the **ilium**, the **ischium**, and the **pubis**. These bones are fused in an adult. The bones of each lower extremity are the **femur, patella, tibia, fibula**, seven **tarsal bones**, five **metatarsal bones**, and fourteen **phalanges**. In addition, two or more sesamoid bones at the interphalangeal joints are usually present.

The various surface features used to identify specific bones are presented in Table 6.1.

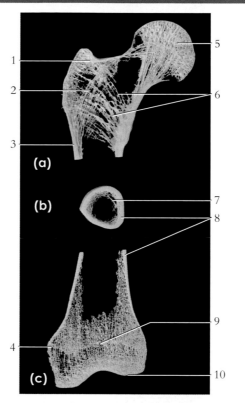

Chapter 6

Figure 6.1 The sections of a human femur showing the outer compact bone and inner spongy bone. Note the trajectorial (stress) lines of the trabeculae of spongy bone.
(a) A coronal section of the proximal end of the femur
(b) A cross section of the body of the femur
(c) A coronal section of the distal end of the femur
1. Greater trochanter
2. Spongy bone
3. Compact bone
4. Femoral epicondyle
5. Head of femur
6. Trabeculae of spongy bone
7. Spongy bone surrounding medullary cavity
8. Compact bone
9. Epiphyseal line (remnant of epiphyseal plate)
10. Condyle of femur

Table 6.1 Surface features used to identify specific regions of bones.

Articulating surfaces	Description	Example
Condyle	Large, rounded, articulating knob	Mandible
Facet	Flattened or shallow articulating surface	Vertebra
Head	Prominent, rounded, articulating end of a bone	Femur
Nonarticulating prominences		
Crest	Narrow, ridgelike projection	Illium
Epicondyle	Projection above a condyle	Humerus
Process	Any marked bony prominence	Radius
Spine	Sharp, slender process	Scapula
Trochanter	Large process found only on the femur	Femur
Tubercle	Small rounded process	Humerus
Tuberosity	Large roughened process	Tibia
Depressions and openings		
Alveolus	Deep pit or socket	Maxilla
Fissure	Narrow, slitlike opening	Sphenoid
Foramen (plural, foramina)	Rounded opening through a bone	Sphenoid
Fossa	Flattened or shallow surface	Scapula
Fovea	Small pit or depression; may or may not be articular	Femur
Meatus, or canal	Tubelike passageway through a bone	Temporal bone
Sinus	Cavity or hollow space in a bone	Maxilla
Sulcus	Groove in a bone	Humerus

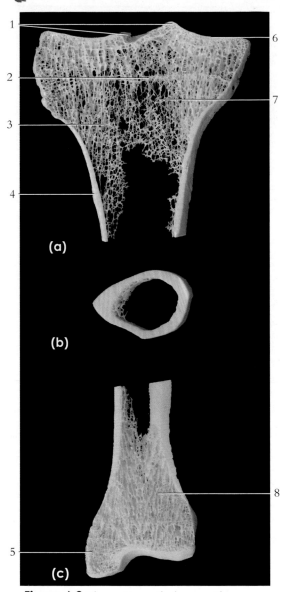

(a)

(b)

(c)

Figure 6.2 The sections of a human tibia showing the outer compact bone and the inner spongy bone. Note the trajectorial (stress) lines of the trabeculae of spongy bone.
(a) A coronal section of the proximal end of the tibia
(b) A cross section of the body of tibia
(c) A coronal section of the distal end of the tibia

1. Intercondylar eminences
2. Epiphyseal line
3. Spongy bone
4. Compact bone
5. Medial malleolus
6. Lateral condyle
7. Trabeculae of spongy bone
8. Spongy bone

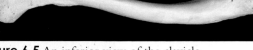

Figure 6.5 An inferior view of the clavicle.
1. Conoid tubercle
2. Acromial extremity
3. Sternal extremity

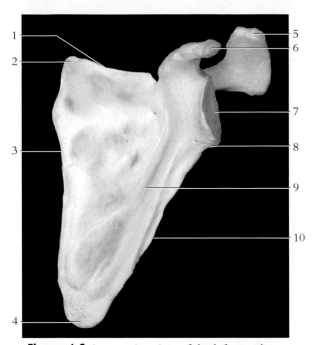

Figure 6.3 An anterior view of the left scapula.
1. Superior border
2. Superior angle
3. Medial (vertebral) border
4. Inferior angle
5. Acromion
6. Coracoid process
7. Glenoid fossa
8. Infraglenoid tubercle
9. Subscapular fossa
10. Lateral (axillary) border

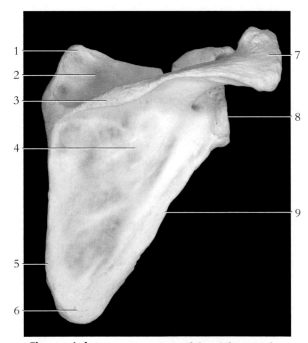

Figure 6.4 A posterior view of the right scapula.
1. Superior angle
2. Supraspinous fossa
3. Spine
4. Infraspinous fossa
5. Medial (vertebral) border
6. Inferior angle
7. Acromion
8. Glenoid fossa
9. Lateral (axillary) border

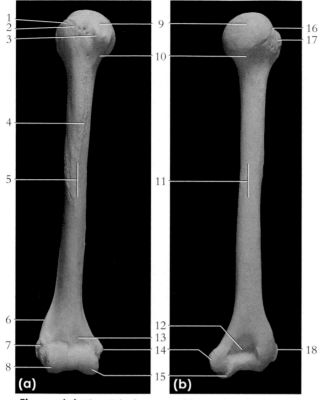

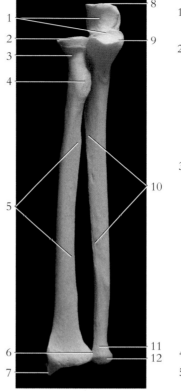

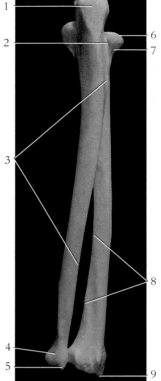

Figure 6.6 The right humerus (a) anterior view (b) posterior view.

1. Greater tubercle
2. Intertubercular groove
3. Lesser tubercle
4. Deltoid tuberosity
5. Anterior body (shaft) of humerus
6. Lateral supracondylar ridge
7. Lateral epicondyle
8. Capitulum
9. Head of humerus
10. Surgical neck
11. Posterior body (shaft) of humerus
12. Olecranon fossa
13. Coronoid fossa
14. Medial epicondyle
15. Trochlea
16. Anatomical neck
17. Greater tubercle
18. Lateral epicondyle

Figure 6.7 An anterior view of the right ulna and radius.

1. Trochlear notch
2. Head of radius
3. Neck of radius
4. Radial tuberosity
5. Interosseous margin
6. Ulnar notch of radius
7. Styloid process of radius
8. Olecranon
9. Coronoid process
10. Interosseous margin
11. Neck of ulna
12. Head of ulna

Figure 6.8 A posterior view of the right ulna and radius.

1. Olecranon
2. Radial notch of ulna
3. Interosseous margin
4. Head of ulna
5. Styloid process of ulna
6. Head of radius
7. Neck of radius
8. Interosseous margin
9. Styloid process of radius

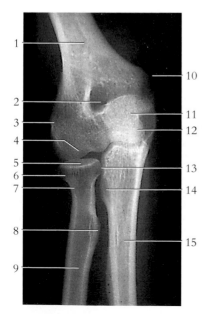

Figure 6.9 A radiograph of the left elbow region, posterior view.

1. Humerus
2. Olecranon fossa of humerus
3. Lateral epicondyle of humerus
4. Capitulum
5. Articular surface of radius
6. Head of radius
7. Neck of radius
8. Radial tuberosity
9. Radius
10. Medial epicondyle of humerus
11. Olecranon
12. Trochlea
13. Radial notch of ulna
14. Tuberosity of ulna
15. Ulna

Figure 6.10 A radiograph showing fractures of the ulna and radius of a ten-year-old child. Notice the distal epiphyseal plates of the ulna and the radius.

1. Ulna
2. Radius
3. Fractures
4. Epiphyseal plates

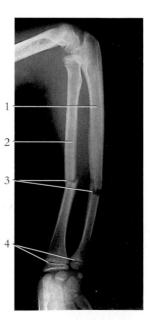

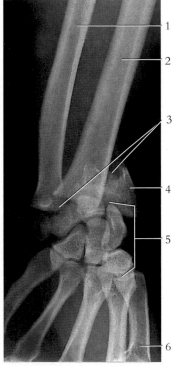

Figure 6.11 A radiograph showing a fracture of the distal portion of the left radius and medial displacement of the antebrachium.
1. Ulna
2. Radius
3. Site of fracture
4. Styloid process of radius
5. Carpal bones
6. First metacarpal bone

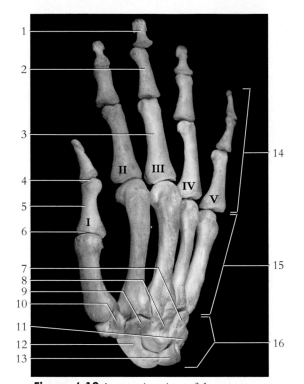

Figure 6.12 A posterior view of the right wrist and hand.
1. Distal phalanx of third digit
2. Middle phalanx of third digit
3. Proximal phalanx of third digit
4. Head of phalanx
5. Body of phalanx
6. Base of phalanx
7. Hamate bone
8. Capitate bone
9. Trapezoid bone
10. Trapezium bone
11. Triquetrum bone
12. Scaphoid bone
13. Lunate bone
14. Phalanges of fifth digit
15. Metacarpal bones (first–fifth)
16. Carpal bones

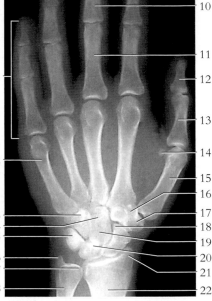

Figure 6.13 A radiograph of the right wrist and hand, posterior view.
1. Phalanges of fifth digit
2. Fifth metacarpal bone
3. Hamate bone
4. Capitate bone
5. Triquetrum and pisiform bones
6. Styloid process of ulna
7. Distal radioulnar joint
8. Ulna
9. Distal phalanx of third digit
10. Middle phalanx of third digit
11. Proximal phalanx of third digit
12. Distal phalanx of pollex (thumb)
13. Proximal phalanx of pollex (thumb)
14. Sesamoid bone
15. First metacarpal bone
16. Trapezoid bone
17. Trapezium bone
18. Scaphoid bone
19. Capitate bone
20. Lunate bone
21. Styloid process of radius
22. Radius

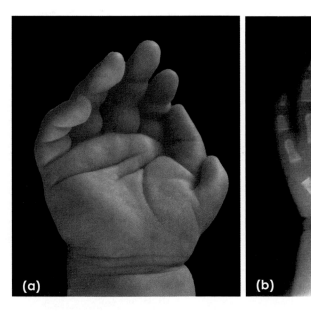

Figure 6.14 A photograph (a) and a radiograph (b) showing polydactyly, having extra digits. Polydactyly is a common congenital deformity of the hand, although it also occurs in the foot. Notice that the carpal bones are still cartilaginous in a newborn and do not show up on a radiograph.

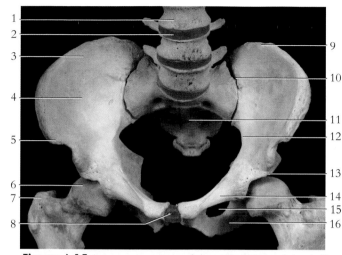

Figure 6.15 An anterior view of the articulated pelvic girdle showing the two coxal bones, the sacrum, and the two femora.

1. Lumbar vertebra
2. Intervertebral disc
3. Ilium
4. Iliac fossa
5. Anterior superior iliac spine
6. Head of femur
7. Greater trochanter
8. Symphysis pubis
9. Crest of the ilium
10. Sacroiliac joint
11. Sacrum
12. Pelvic brim
13. Acetabulum
14. Pubic crest
15. Obturator foramen
16. Ischium

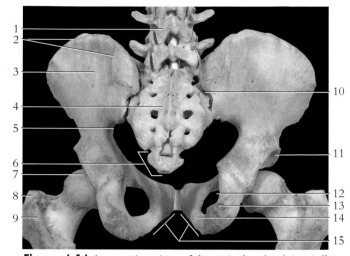

Figure 6.16 A posterior view of the articulated pelvic girdle showing the two coxal bones, the sacrum, and the two femora.

1. Lumbar vertebra
2. Crest of ilium
3. Ilium
4. Sacrum
5. Greater sciatic notch
6. Coccyx
7. Head of femur
8. Greater trochanter
9. Intertrochanteric crest
10. Sacroiliac joint
11. Acetabulum
12. Obturator foramen
13. Ischium
14. Pubis
15. Pubic angle

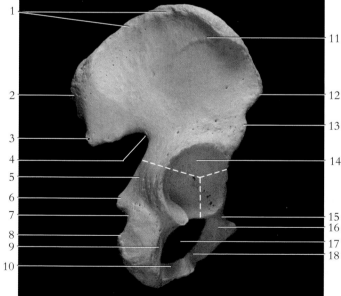

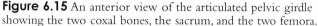

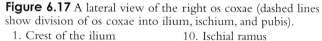

Figure 6.17 A lateral view of the right os coxae (dashed lines show division of os coxae into ilium, ischium, and pubis).

1. Crest of the ilium
2. Posterior superior iliac spine
3. Posterior inferior iliac spine
4. Greater sciatic notch
5. Ischial body
6. Ischial spine
7. Lesser sciatic notch
8. Ischial tuberosity
9. Ischium
10. Ischial ramus
11. Ilium
12. Anterior superior iliac spine
13. Anterior inferior iliac spine
14. Acetabulum
15. Superior ramus of pubis
16. Pubis
17. Obturator foramen
18. Inferior ramus of pubis

Figure 6.18 A medial view of the right os coxae (dashed lines show division of os coxae into ilium, ischium, and pubis).

1. Ilium
2. Iliac fossa
3. Anterior superior iliac spine
4. Anterior inferior iliac spine
5. Superior ramus of pubis
6. Pubic crest
7. Pubis
8. Inferior pubic ramus
9. Iliac tuberosity
10. Posterior superior iliac spine
11. Auricular surface
12. Posterior inferior iliac spine
13. Greater sciatic notch
14. Ischial spine
15. Ischium
16. Lesser sciatic notch
17. Obturator foramen
18. Ischial ramus

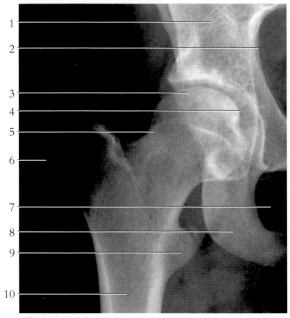

Figure 6.19 A radiograph of the right os coxae and femur, anterior view.

1. Ilium
2. Pelvic brim
3. Head of femur
4. Acetabulum
5. Neck of femur
6. Greater trochanter
7. Obturator foramen
8. Ischium
9. Lesser trochanter
10. Femur

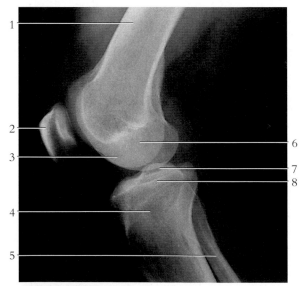

Figure 6.20 A radiograph of the right knee region, medial view (flexed).

1. Femur
2. Patella
3. Medial condyle of femur
4. Tibia
5. Fibula
6. Medial epicondyle of femur
7. Intercondylar eminences of tibia
8. Medial condyle of tibia

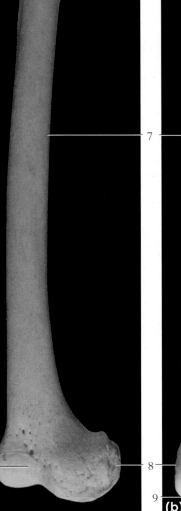

Figure 6.21 The right femur (a) anterior view (b) posterior view.

1. Fovea capitis femoris
2. Intertrochanteric line
3. Patellar surface
4. Head of femur
5. Neck of femur
6. Lesser trochanter
7. Body (shaft) of femur
8. Medial epicondyle
9. Medial condyle
10. Greater trochanter
11. Intertrochanteric crest
12. Linea aspera on body (shaft) of femur
13. Lateral epicondyle
14. Lateral condyle
15. Intercondylar fossa

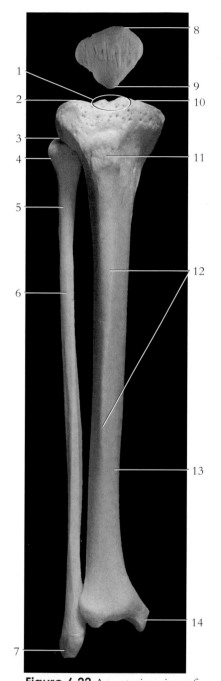

Figure 6.22 An anterior view of the right patella, tibia, and fibula.
1. Intercondylar tubercles
2. Lateral condyle
3. Tibial articular facet of fibula
4. Head of fibula
5. Neck of fibula
6. Body (shaft) of fibula
7. Lateral malleolus
8. Base of patella
9. Apex of patella
10. Medial condyle
11. Tibial tuberosity
12. Anterior crest of tibia
13. Body (shaft) of tibia
14. Medial malleolus

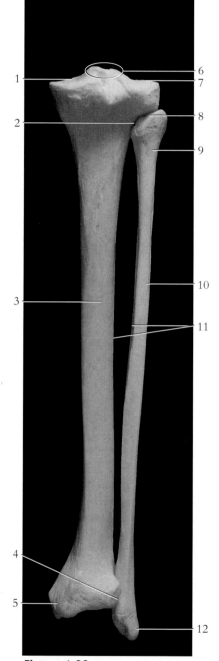

Figure 6.23 A posterior view of the right tibia and fibula.
1. Medial condyle of tibia
2. Fibular articular facet of tibia
3. Body (shaft) of tibia
4. Fibular notch of tibia
5. Medial malleolus
6. Intercondylar tubercles
7. Lateral condyle of tibia
8. Head of fibula
9. Neck of fibula
10. Body (shaft) of fibula
11. Interosseous border
12. Lateral malleolus

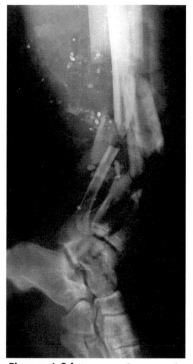

Figure 6.24 A severe fracture of the leg and ankle (talus bone). In this patient, the trauma was so extensive amputation of the leg was necessary.

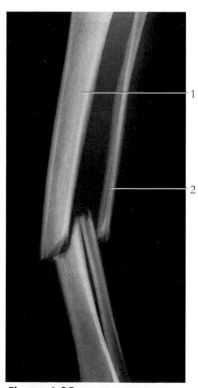

Figure 6.25 A radiograph showing fractures of the tibia and fibula.
1. Tibia
2. Fibula

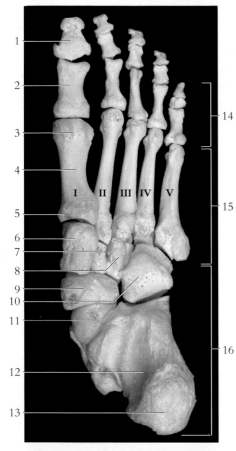

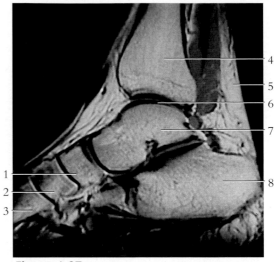

Figure 6.26 An inferior view of the right foot.
1. Distal phalanx of first digit
2. Proximal phalanx of first digit
3. Head of first metatarsal bone
4. Body of first metatarsal bone
5. Base of first metatarsal bone
6. Medial cuneiform
7. Intermediate cuneiform
8. Lateral cuneiform
9. Navicular bone
10. Cuboid bone
11. Talus bone
12. Calcaneus bone
13. Tuberosity of calcaneus
14. Phalanges of fifth digit
15. Metatarsal bones (first–fifth)
16. Tarsal bones

Figure 6.27 An MR image of the ankle region, medial view.
1. Navicular bone
2. Medial cuneiform
3. First metatarsal bone
4. Tibia
5. Tendo calcaneus
6. Tibiotalar joint
7. Talus
8. Calcaneus

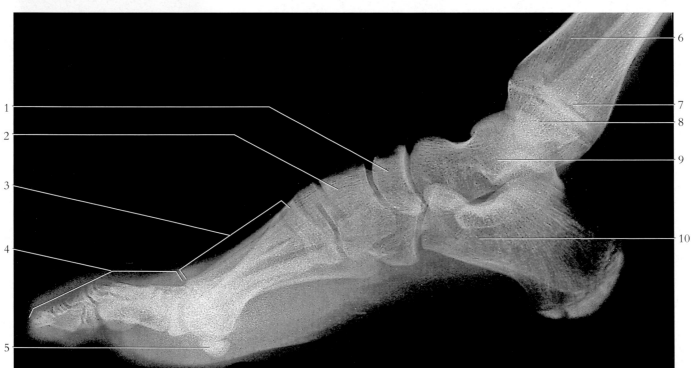

Figure 6.28 A radiograph of the right foot, medial view.
1. Navicular bone
2. First (medial) cuneiform
3. Metatarsal bones
4. Phalanges
5. Sesamoid bone
6. Tibia
7. Fibula (superimposed)
8. Tibiotalar joint
9. Talus
10. Calcaneus

Skeletal System: Articulations

Chapter 7

A joint is a region where bones articulate. The structure of a joint determines its range of movement. Not all joints are flexible, however, and as one part of the body moves, other joints remain rigid to stabilize the body and maintain balance. **Arthrology** is the study of joints and **kinesiology** is the study of body movement, or the functional relationship between the skeleton, joints, muscles, and innervation (nerve supply) as they work together to produce coordinated movement.

The joints of the body are structurally classified into three basic kinds:

1. **Fibrous joints**

   Fibrous connective tissues join the skeletal structures; the joints lack a joint capsule, but in some slight movement is possible.

2. **Cartilaginous joints**

   Fibrocartilage or hyaline cartilage joins the skeletal structures; the joints lack a joint capsule, but in some, slight movement is possible.

3. **Synovial joints**

   Joint capsules containing synovial fluid are present between the articulating bones; articular cartilages and ligaments supporting the articulating bones are also present, which permit freedom of movement. Synovial joints are the freely moveable joints of the body.

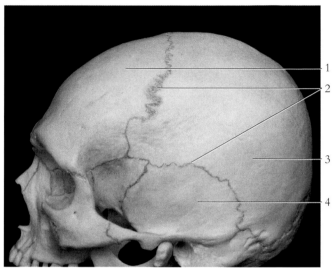

Figure 7.1 A Suture located between cranial bones is a type of fibrous joint.

| 1. Frontal bone | 3. Parietal bone |
| 2. Sutures | 4. Temporal bone |

Table 7.1 Movements permitted at synovial joints.

Type of movement	Description
Angular movement	Increase or decrease the joint angle
Flexion	Decreasing the angle between two bones
Extension	Increasing the angle between two bones
Hyperextension	Excessive extension beyond 180° (angle of anatomical position)
Dorsiflexion	Bending the foot toward the tibia
Plantar flexion	Bending the foot away from the tibia
Abduction	Movement of a body part away from the axis of the body, or away from the midsagittal plane, in a lateral direction
Adduction	Movement of a body part toward the axis of the body, or toward the midsagittal plane, in a mesial direction
Inversion	Movement of the sole of the foot inward, or medially
Eversion	Movement of the sole of the foot outward, or laterally
Opposition	Movement of the thumb toward the surface of the palm
Protraction	Moving a part of the body anteriorly in the horizontal plane
Retraction	Moving a part of the body posteriorly in the horizontal plane
Elevation	Moving a part of the body in the superior direction
Depression	Moving a part of the body in the inferior direction
Circular movement	
Rotation	Turning of a bone at joint axis
Pronation	Rotation of the forearm causing the wrist and hand to go from palm facing front to palm facing back
Supination	Rotation of the forearm causing the wrist and hand to go from palm facing back to palm facing front
Circumduction	Movement of a body segment in a circular, conelike motion

Table 7.2 Types of joints.

Type	Structure and movement	Example
Fibrous	Fibrous connective tissue joins the skeletal structures	
Suture	Frequently serrated edges of articulating bones, separated by a thin layer of fibrous tissue; no movement	Joints between cranial bones
Syndesmosis	Articulating bones bound by an interosseous ligament; slight movement	Joints between tibia-fibula and radius-ulna
Gomphosis	Periodontal ligament binding teeth into dental aveoli of bone; no movement	Teeth secured in the dental alveoli (teeth sockets)
Cartilaginous	Fibrocartilage or hyaline cartilage joins the skeletal structures	
Symphysis	Thin pad of fibrocartilage between articulating bones; slight movement	Symphysis pubis, and intervertebral joints (discs)
Synchondrosis	Hyaline cartilage between skeletal structures; no movement	Epiphyseal plates between diaphysis and epiphyses of long bones; ribs and costal cartilages
Synovial	Joint capsule between articulating bones, containing synovial fluid; extensive movement often possible	
Gliding	Flattened or slightly curved articulating surfaces; sliding movement	Intercarpal and intertarsal joints
Hinge	Concave surface of one bone articulates with a depression of another; bending motion in one plane	Humeroulnar (elbow) and tibiofemoral (knee) joints; inter-phalangeal (finger and toe) joints of digits
Pivot	Conical surface of one bone articulates with a depression of another; rotation about a central axis; rotational movement	Atlantoaxial joint; proximal radioulnar joint
Condyloid	Oval condyle of one bone articulates with elliptical cavity of another; biaxial movement	Radiocarpal (wrist) joint
Saddle	Concave and convex surface on each articulating bone; wide range of movement; biaxial movement	Carpometacarpal joint at the base of the thumb
Ball-and-socket	Rounded convex surface of one bone articulates with cuplike socket of another; movement in all planes and rotation	Glenohumeral (shoulder) and coxal (hip) joints

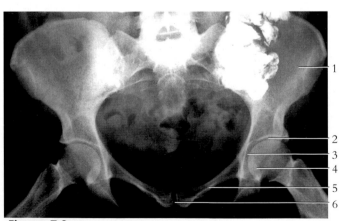

Figure 7.2 A symphysis is a type of cartilaginous joint. The symphysis pubis can be seen in this radiograph of the pelvic girdle. Note the head of the femur articulating with the acetabulum of the os coxae forming a ball-and-socket joint is a freely moveable synovial joint.

1. Ilium
2. Acetabulum
3. Ball-and-socket joint
4. Head of femur
5. Pubis
6. Symphysis pubis

Figure 7.3 An intervertebral joint between vertebral bodies is a type of cartilaginous joint. The intervertebral discs of intervertebral joints can be readily seen in a lateral MR image of the lower back.

1. Spinal cord
2. Body of second lumbar vertebra (L2)
3. Intervertebral disc

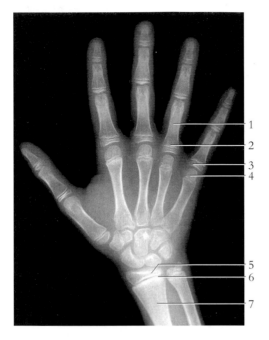

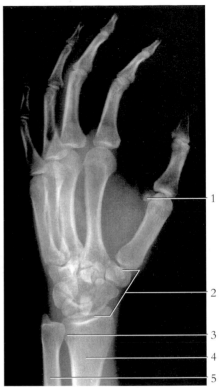

Figure 7.4 A synchondrosis is a type of cartilaginous joint located within a long bone at both the proximal and the distal epiphyseal plates as seen in a radiograph of a child's hand. Mitotic activity at a synchondrotic joint is responsible for linear (length) bone growth.

1. Diaphysis of phalanx
2. Epiphyseal plate
3. Proximal epiphysis of phalanx
4. Distal epiphysis of metacarpal bone
5. Distal epiphysis of radius
6. Epiphyseal plate
7. Diaphysis of radius

Figure 7.5 The side-to-side articulation of the ulna and radius forms a syndesmosis that is tightly bound by an interosseous ligament (not seen in radiograph). A syndesmosis is a type of fibrous joint that permits slight movement.

1. Sesamoid bone   4. Radius
2. Carpal bones   5. Ulna
3. Syndesmosis

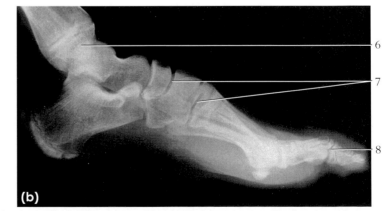

(a)   (b)

Figure 7.6 Radiographs of the (a) wrist and hand and (b) the ankle and foot. Several kinds of synovial joints (freely moveable joints) are contained in these regions.

1. Hinge joints—interphalangeal joints of the second digit (index finger)
2. Condyloid joint—metacarpophalangeal joint of the second digit (index finger)
3. Saddle joint—carpometacarpal joint of the pollex (thumb)
4. Gliding joint—intercarpal joint between capitate and scaphoid bones
5. Condyloid joint—radiocarpal joint
6. Condyloid joint at the ankle
7. Gliding joints between tarsal bones
8. Hinge joint between phalanges of digit

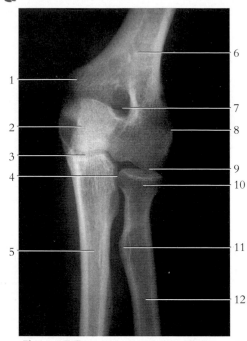

Figure 7.7 A radiograph of the elbow region depicting two types of synovial joints. The humeroulnar joint is a hinge joint that permits movement along a single plane. The proximal radioulnar and humeroradial joints are both pivot joints that permit rotational movement.

1. Medial epicondyle
2. Olecranon
3. Humeroulnar joint
4. Proximal radioulnar joint
5. Ulna
6. Humerus
7. Olecranon fossa
8. Lateral epicondyle
9. Humeroradial joint
10. Head of radius
11. Radial tuberosity
12. Radius

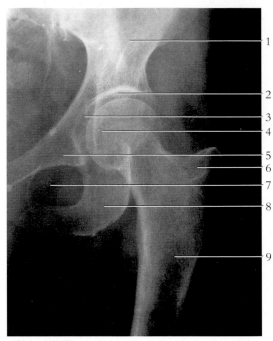

Figure 7.8 The ball-and-socket joint seen in this radiograph of the hip is a synovial joint. The coxal joint (hip joint) is formed by the head of the femur articulating with the acetabulum of the os coxae.

1. Ilium
2. Acetabulum
3. Ball-and-socket joint
4. Head of femur
5. Pubis
6. Greater trochanter
7. Obturator foramen
8. Ischium
9. Femur

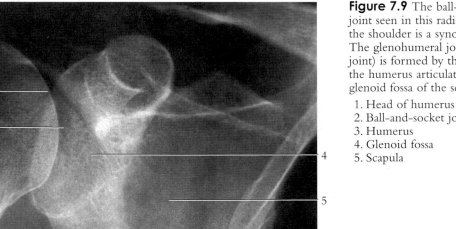

Figure 7.9 The ball-and-socket joint seen in this radiograph of the shoulder is a synovial joint. The glenohumeral joint (shoulder joint) is formed by the head of the humerus articulating with the glenoid fossa of the scapula.

1. Head of humerus
2. Ball-and-socket joint
3. Humerus
4. Glenoid fossa
5. Scapula

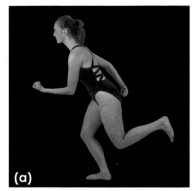

(a) Flexion at left shoulder, elbow, and knee joints. Extension of right shoulder, and elbow.

(b) Maximal flexion at each of the principal body joints.

(c) Rotation at the joints of the neck; elevation at the shoulder joint; and flexion at the right elbow and right wrist joints.

(d) Abduction at the shoulder and hip joints. Also, abduction at the metacarpophalangeal joints at the base of the digits.

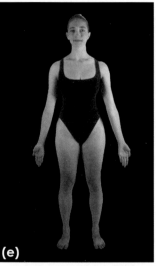

(e) Adduction at the shoulder and hip joints. Also, adduction at the joints of the fingers and toes.

(f) Lateral (right and left) rotation at joints of the vertebral column.

(g) Lateral bending (flexion) at joints of the vertebral column.

(h) Flexion at the joints of the vertebral column.

(i) Hyperextension at the joints of the vertebral column.

(j) Flexion at the shoulder, hip, and knee joints on left side of body; extension at the elbow and wrist joints; knee extension on right side of body.

(k) Hyperextension at the shoulder and hip joints on left side of body. Plantar flexion at the left ankle joint.

Figure 7.10 A photograghic summary of joint movements (a-k).

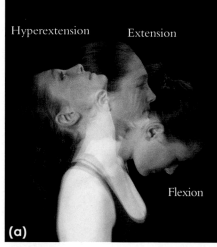

(a)

Hyperextension, extension, and flexion at the cervical intervertebral joints of the neck.

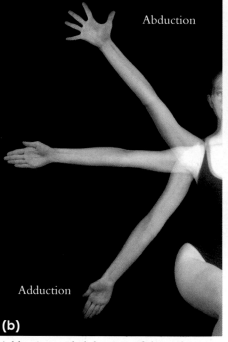

(b)

Adduction and abduction of the right arm at the shoulder joint and fingers at the metacarpophalangeal joints.

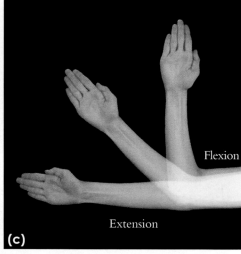

(c)

Flexion and extension at the right elbow joint.

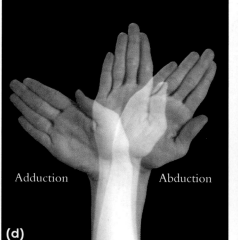

(d)

Abduction and adduction of the right hand at the wrist.

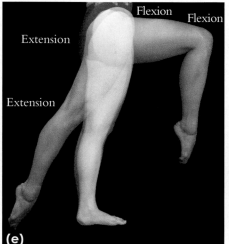

(e)

Flexion and extension at the right hip and knee joints.

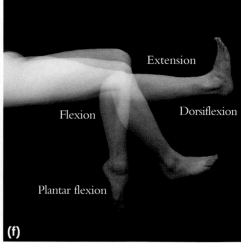

(f)

Flexion and extension at the knee joint and plantar flexion and dorsiflexion at the ankle joint.

Figure 7.11 A visualization of angular movements permitted at synovial joints (a–f).

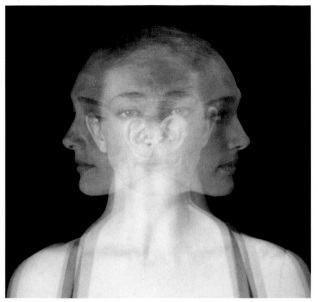

Figure 7.12 Rotation of the head at the cervical vertebrae–principally at the atlantoaxial joint.

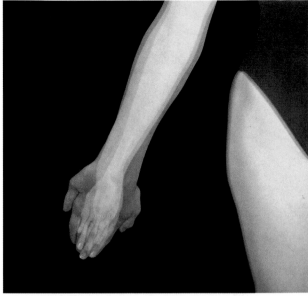

Figure 7.13 Rotation of the antebrachium (forearm) at the proximal radioulnar joint–an example of pronation and supination.

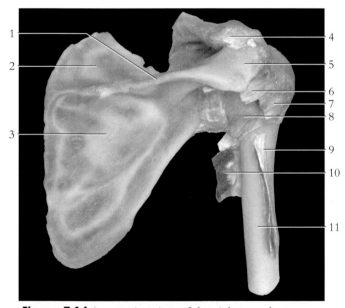

Figure 7.14 A posterior view of the right scapula, proximal humerus, and glenohumeral (shoulder) joint.
1. Spine of scapula
2. Supraspinous fossa
3. Infraspinous fossa
4. Acromioclavicular ligament
5. Acromion
6. Tendon of infraspinatus muscle
7. Tendon of teres minor muscle
8. Articular capsule of glenohumeral joint
9. Origin of lateral head of triceps brachii muscle
10. Teres major muscle
11. Humerus

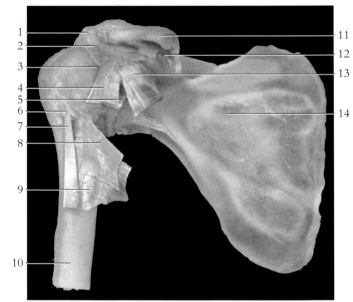

Figure 7.15 An anterior view of the right scapula, proximal humerus, and glenohumeral (shoulder) joint.
1. Acromion
2. Coracoacrominal ligament
3. Articular capsule of glenohumeral joint
4. Tendon of short head of biceps brachii muscle
5. Tendon of coracobrachialis muscle
6. Tendon sheath
7. Tendon of long head of biceps brachii muscle
8. Tendon of latissimus dorsi muscle
9. Tendon of teres major muscle
10. Humerus
11. Clavicle
12. Subclavius muscle
13. Tendon of pectoralis minor muscle
14. Subscapular fossa

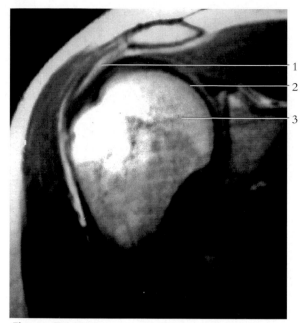

Figure 7.16 The structure of the humeral joint (shoulder joint) as seen in a MR image.
1. Subdeltoid bursa
2. Articular cavity of shoulder joint
3. Head of humerus

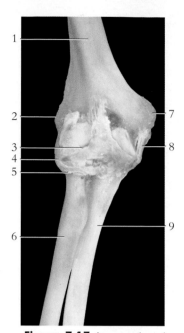

Figure 7.17 An anterior view of the right elbow region.
1. Humerus
2. Lateral epicondyle of humerus
3. Articular capsule of elbow joint
4. Annular ligament
5. Head of radius
6. Radius
7. Medial epicondyle
8. Ulnar collateral ligament
9. Ulna

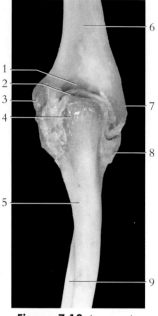

Figure 7.18 A posterior view of the right elbow region.
1. Olecranon fossa
2. Articular capsule of elbow joint
3. Medial epicondyle
4. Olecranon
5. Ulna
6. Humerus
7. Lateral epicondyle
8. Head of radius
9. Radius

Figure 7.19 The structure of the elbow joint as seen in a radiograph.
1. Articular cavity of elbow joint
2. Head of radius
3. Neck of radius
4. Radial tuberosity
5. Radius
6. Humerus
7. Olecranon
8. Ulna

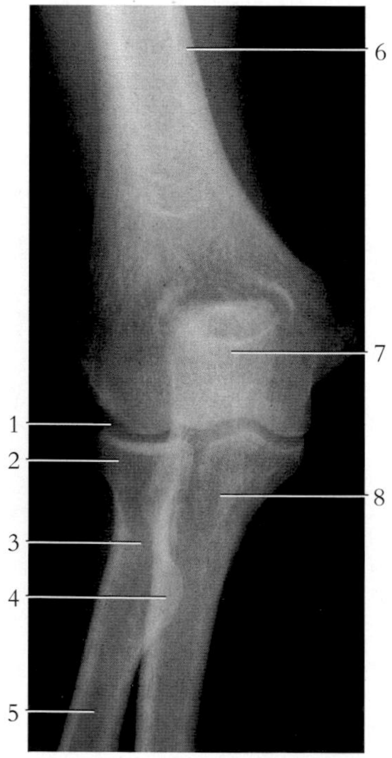

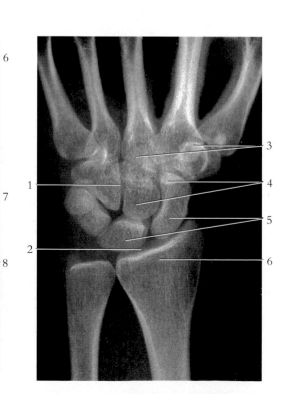

Figure 7.20 The structure of the wrist joint as seen in a MR image.
1. Gliding joint
2. Condyloid joint
3. Heads of second and third metacarpal bones
4. Distal carpal bones (trapezoid and capitate bones)
5. Proximal carpal bones (scaphoid and lunate bones)
6. Distal epiphysis of radius

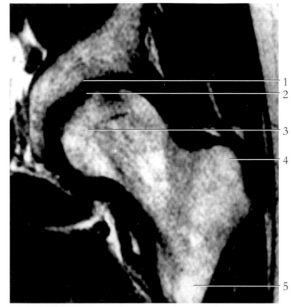

Figure 7.21 The structure of the coxal joint (hip joint) as seen in a MR image.

1. Acetabulum
2. Articular cavity of hip joint
3. Head of femur
4. Greater trochanter of femur
5. Body of femur

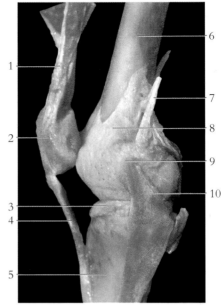

Figure 7.22 A medial view of the right tibiofemoral (knee) joint

1. Tendon of quadriceps femoris muscle
2. Patella
3. Medial meniscus
4. Patellar ligament
5. Tibia
6. Femur
7. Tendon of adductor magnus muscle
8. Articular capsule of tibiofemoral (knee) joint
9. Medial epicondyle of femur
10. Tibial collateral ligament

Figure 7.23 A lateral view of right tibio-femoral (knee) joint.

1. Bursa
2. Lateral collateral ligament
3. Articular cartilage
4. Proximal epiphysis
5. Tibiofibular joint
6. Femur
7. Articular capsule
8. Lateral meniscus
9. Patellar ligament
10. Patella

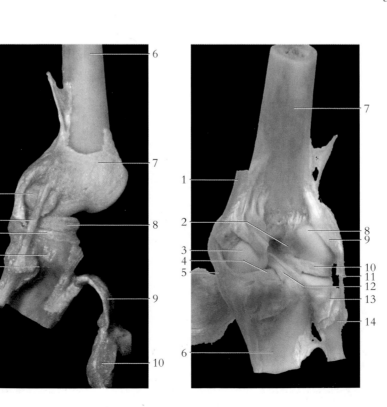

Figure 7.24 A posterior view of right tibiofemoral (knee) joint.

1. Tendon of adductor magnus muscle
2. Anterior cruciate ligament
3. Medial femoral condyle
4. Medial meniscus
5. Medial collateral ligament
6. Tibia
7. Femur
8. Lateral femoral condyle
9. Tendon of popliteus muscle
10. Lateral meniscus
11. Lateral collateral ligament
12. Posterior cruciate ligament
13. Proximal tibiofibular joint
14. Head of fibula

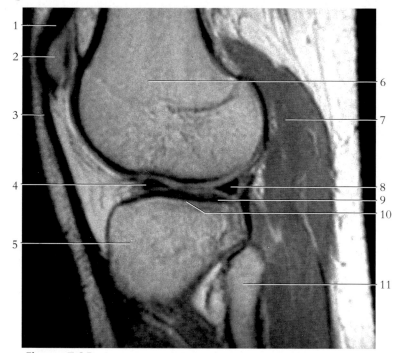

Figure 7.25 A sagittal MR image of tibiofemoral joint (knee joint).
1. Tendon of quadriceps muscle
2. Patella
3. Patellar ligament
4. Lateral meniscus, anterior horn
5. Head of tibia
6. Femur
7. Gastrocnemius muscle
8. Lateral meniscus, posterior horn
9. Synovial fluid
10. Articular cartilage
11. Head of fibula

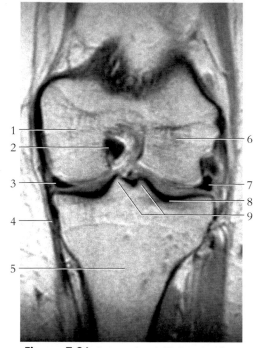

Figure 7.26 A posterior MR image of tibiofemoral joint (knee joint).
1. Medial condyle
2. Insertion of posterior cruciate ligament
3. Medial meniscus
4. Tibial collateral ligament
5. Tibia
6. Lateral condyle of femur
7. Lateral meniscus
8. Synovial fluid
9. Intercondylar eminences

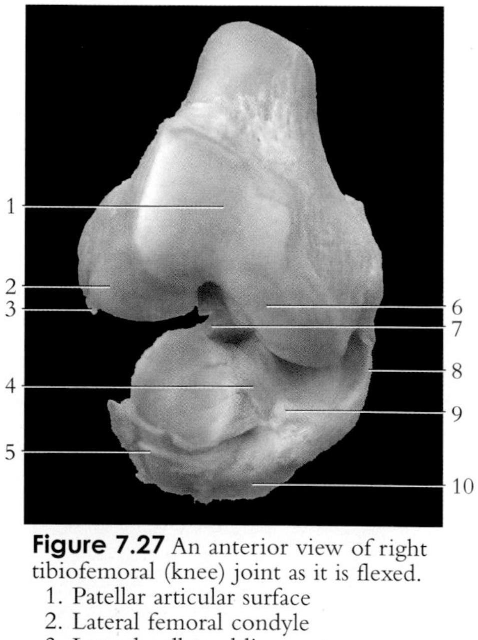

Figure 7.27 An anterior view of right tibiofemoral (knee) joint as it is flexed.
1. Patellar articular surface
2. Lateral femoral condyle
3. Lateral collateral ligament
4. Anterior cruciate ligament
5. Articular cartilage of tibia
6. Medial femoral condyle
7. Posterior cruciate ligament
8. Medial collateral ligament
9. Medial meniscus
10. Tibia

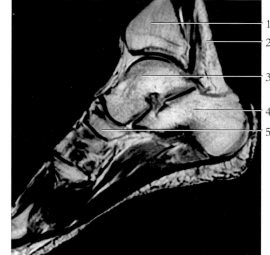

Figure 7.28 A sagittal MR image of the talocrual (ankle) and tarsal joints of the foot.
1. Tibia
2. Tendo calcaneus
3. Talus
4. Calcaneus
5. Navicular bone

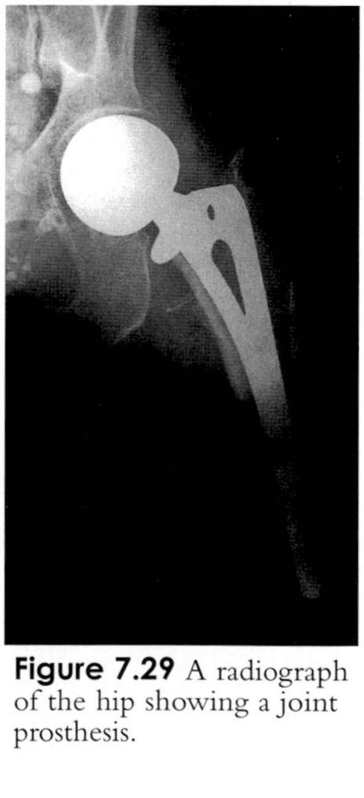

Figure 7.29 A radiograph of the hip showing a joint prosthesis.

Muscular System

Most skeletal muscles span joints and are attached to a bone at both ends by a **tendon**. Certain tendons, especially at the wrist and ankle, are enclosed individually by **tendon sheaths**. A group of tendons in these areas is also covered by a **retinaculum**. The **origin** of a muscle is the more stationary attachment, and the **insertion** is the more moveable attachment (Fig. 8.1). **Synergistic muscles** contract together. **Antagonistic muscles** perform in opposition to a synergistic group of muscles.

Fascia is a fibrous connective tissue that covers muscle and attaches to the skin. **Superficial fascia** secures the skin to the underlying muscle. **Deep fascia** is an inward extension of the superficial fascia and surrounds adjacent muscles, compartmentalizing and binding them into functional groups.

A skeletal muscle fiber is an elongated, multinuclcated, striated cell (Fig. 8.2). Each fiber is surrounded by a cell membrane, called a **sarcolemma**, and the cytoplasm within the cell is called **sarcoplasm**. Each fiber contains many parallel, thread-like structures called **myofibrils**. Each myofibril is composed of smaller strands called **myofilaments** that contain the contractile proteins **actin** and **myosin**. The regular spatial organization of the contractile proteins within the myofibrils forms the cross-banding striations characteristic of skeletal muscle. A network of membranous compartments, called the **sarcoplasmic reticulum**, extends throughout the cytoplasm.

The myofibrils of a skeletal muscle fiber are arranged into subunits called **sarcomeres**. The dark and light striations of the sarcomere are due to the arrangement of the thick (myosin) and thin (actin) filaments. The dark bands are called **A bands** and the lighter bands, containing only actin filaments, are called **I bands**.

The outer regions of the A bands contain actin and myosin; however, the lighter central regions **(H–zones)** of the A bands contain only myosin. The I bands are bisected by dark **Z lines** where the actin filaments of adjacent sarcomeres join. Muscle fiber contraction results from the interaction of contractile proteins (actin and myosin myofilaments) in which the length of the sarcomeres is reduced.

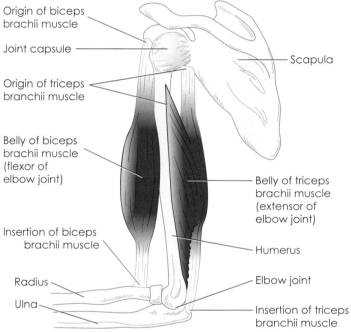

Figure 8.1 The skeletomuscular relationship.

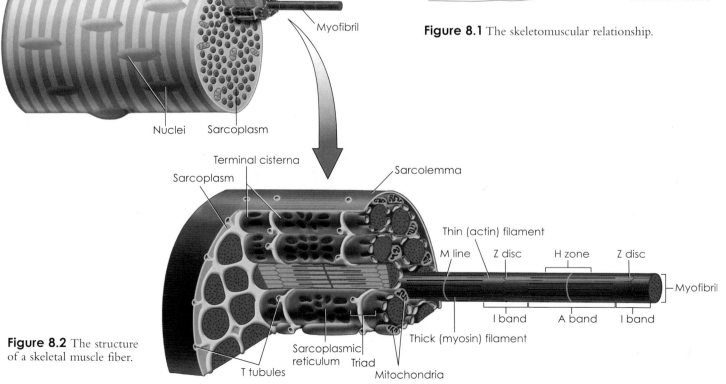

Figure 8.2 The structure of a skeletal muscle fiber.

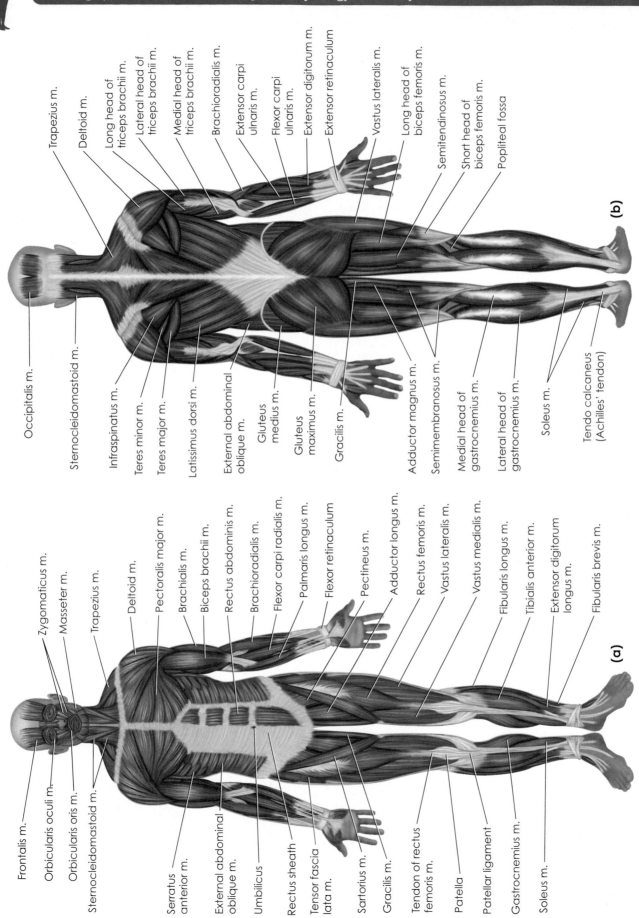

Figure 8.3 An overview of human musculature (a) anterior and (b) posterior (m.=muscle).

Occipitalis m.

Sternocleidomastoid m.

Trapezius m.

Deltoid m.

Long head of triceps brachii m.

Lateral head of triceps brachii m.

Medial head of triceps brachii m.

Brachioradialis m.

Extensor carpi ulnaris m.

Flexor carpi ulnaris m.

Extensor digitorum m.

Extensor retinaculum

Vastus lateralis m.

Long head of biceps femoris m.

Semitendinosus m.

Short head of biceps femoris m.

Popliteal fossa

Infraspinatus m.

Teres minor m.

Teres major m.

Latissimus dorsi m.

External abdominal oblique m.

Gluteus medius m.

Gluteus maximus m.

Gracilis m.

Adductor magnus m.

Semimembranosus m.

Medial head of gastrocnemius m.

Lateral head of gastrocnemius m.

Soleus m.

Tendo calcaneus (Achilles' tendon)

(b)

Frontalis m.

Orbicularis oculi m.

Orbicularis oris m.

Sternocleidomastoid m.

Zygomaticus m.

Masseter m.

Trapezius m.

Deltoid m.

Pectoralis major m.

Brachialis m.

Biceps brachii m.

Rectus abdominis m.

Brachioradialis m.

Flexor carpi radialis m.

Palmaris longus m.

Flexor retinaculum

Pectineus m.

Adductor longus m.

Rectus femoris m.

Vastus lateralis m.

Vastus medialis m.

Fibularis longus m.

Tibialis anterior m.

Extensor digitorum longus m.

Fibularis brevis m.

Serratus anterior m.

External abdominal oblique m.

Umbilicus

Rectus sheath

Tensor fascia lata m.

Sartorius m.

Gracilis m.

Tendon of rectus femoris m.

Patella

Patellar ligament

Gastrocnemius m.

Soleus m.

(a)

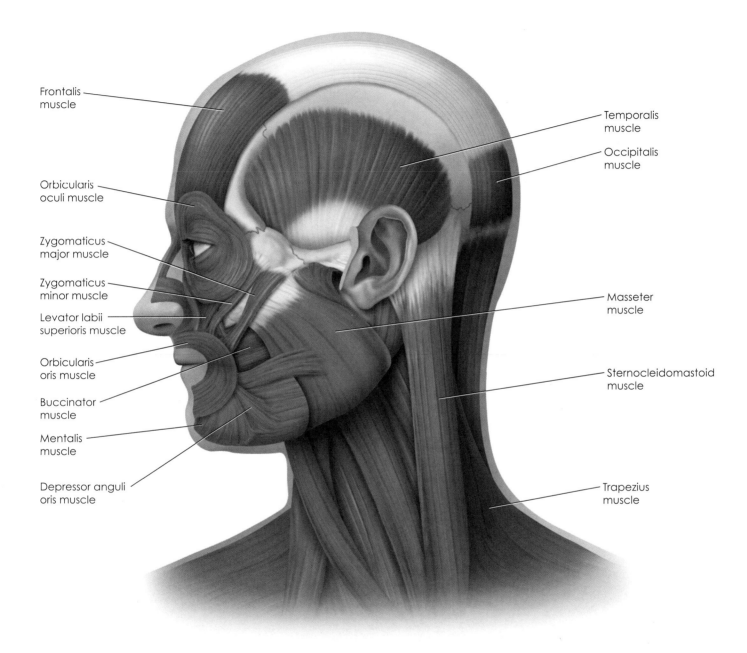

Frontalis
muscle

Orbicularis
oculi muscle

Zygomaticus
major muscle

Zygomaticus
minor muscle

Levator labii
superioris muscle

Orbicularis
oris muscle

Buccinator
muscle

Mentalis
muscle

Depressor anguli
oris muscle

Temporalis
muscle

Occipitalis
muscle

Masseter
muscle

Sternocleidomastoid
muscle

Trapezius
muscle

Figure 8.4 A lateral view of the major muscles of the scalp, face, and neck.

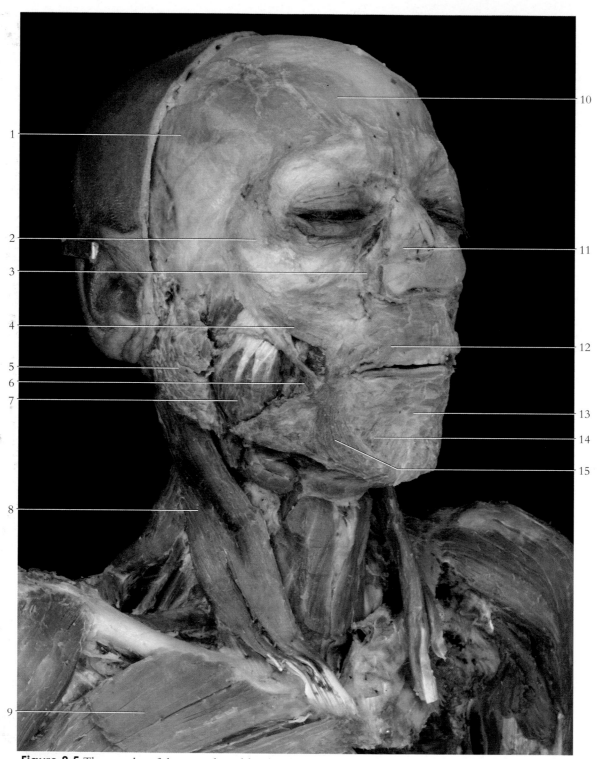

Figure 8.5 The muscles of the anterolateral head and neck regions.

1. Temporalis m.
2. Orbicularis oculi m.
3. Levator labii superioris m.
4. Zygomaticus major m.
5. Parotid gland
6. Buccinator m.
7. Masseter m.
8. Sternocleidomastoid m.
9. Pectoralis major m.
10. Frontalis m.
11. Nasalis m.
12. Orbicularis oris m.
13. Mentalis m.
14. Depressor labii inferioris m.
15. Depressor anguli oris m.

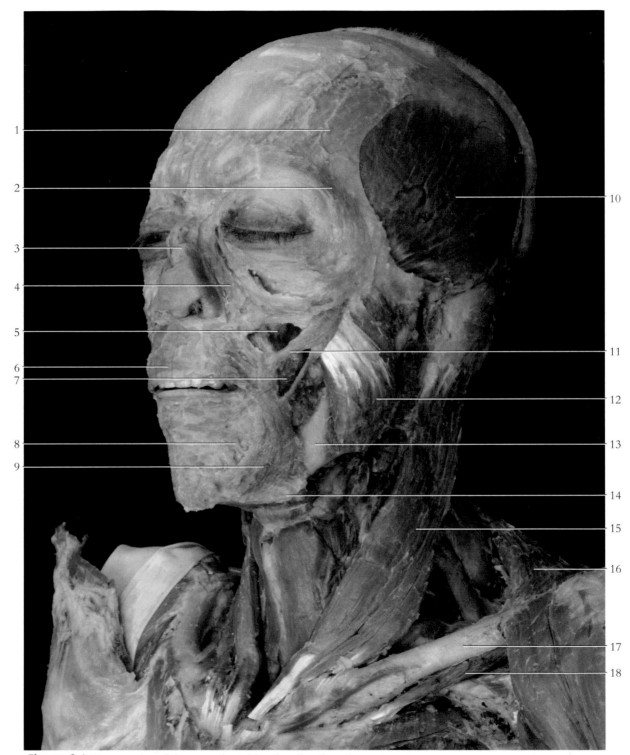

Figure 8.6 The muscles of the anterolateral head and neck regions.

1. Frontalis m.
2. Orbicularis oculi m.
3. Nasalis m.
4. Levator labii superioris m.
5. Levator anguli oris m.
6. Orbicularis oris m.
7. Buccinator m.
8. Depressor labii inferioris m.
9. Depressor anguli oris m.
10. Temporalis m.
11. Zygomaticus major m.
12. Masseter m.
13. Mandible
14. Platysma m. (cut)
15. Sternocleidomastoid m.
16. Trapezius m.
17. Clavicle
18. Subclavius m.

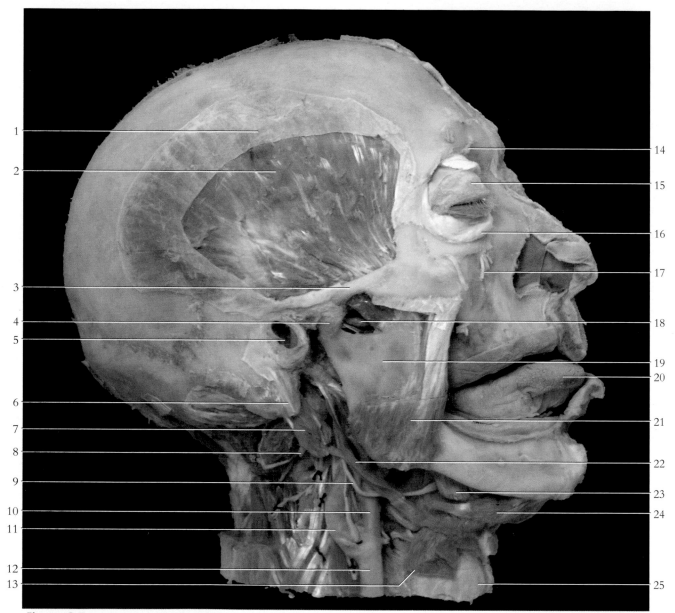

Figure 8.7 A lateral view of the deep structures of the head and neck.

1. Temporal fascia
2. Temporalis m.
3. Zygomatic arch
4. Joint capsule of temporomandibular joint
5. External acoustic canal
6. Mastoid process of temporal bone
7. Posterior belly of digastric m.
8. Vagus nerve (cut)
9. Hypoglossal nerve
10. External cartoid artery
11. Internal cartoid artery
12. Common carotid artery
13. Omohyoid m. (cut)
14. Supraorbital nerve
15. Superior tarsal plate
16. Palpebral fascia
17. Infraorbital nerve
18. Lateral pterygoid m.
19. Mandible
20. Tongue
21. Masseter m. (cut)
22. Stylohyoid m.
23. Mylohyoid m.
24. Anterior belly of digastric muscle
25. Thyroid cartilage of larynx

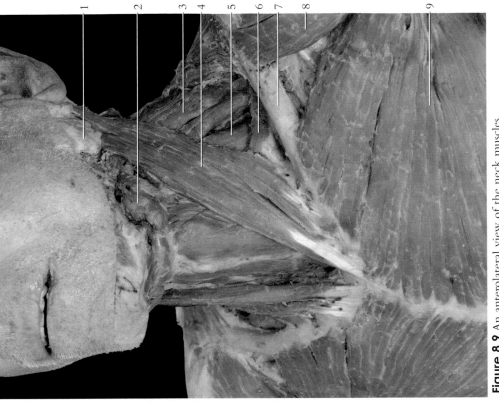

Figure 8.9 An anterolateral view of the neck muscles.
1. Parotid gland
2. Submandibular gland
3. Levator scapulae m.
4. Sternocleidomastoid m.
5. Middle scalene m.
6. Omohyoid m.
7. Clavicle
8. Deltoid m.
9. Pectoralis major m.

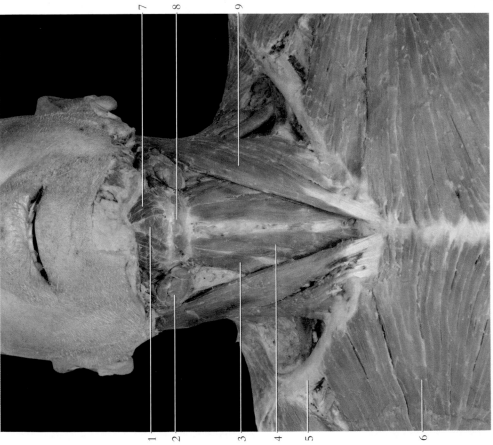

Figure 8.8 An anterior view of the neck muscles.
1. Mylohyoid m.
2. Submandibular gland
3. Omohyoid m.
4. Sternohyoid m.
5. Clavicle
6. Pectoralis major m.
7. Digastric m. (anterior belly)
8. Hyoid bone
9. Sternocleidomastoid m.

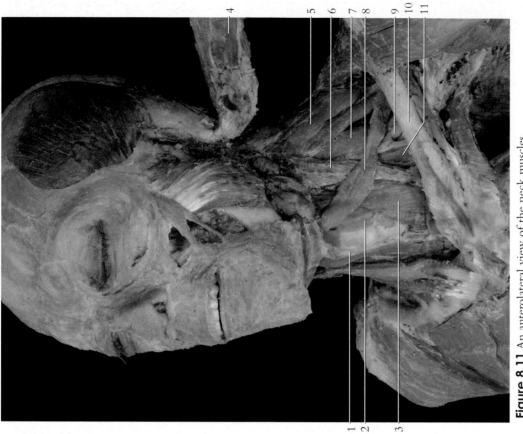

Figure 8.11 An anterolateral view of the neck muscles.

1. Thyroid cartilage
2. Thyrohyoid m.
3. Sternothyroid m.
4. Sternocleidomastoid m. (cut and reflected)
5. Levator scapulae m.
6. Internal jugular vien
7. Middle scalene m.
8. Omohyoid m.
9. Brachial plexus
10. Clavicle
11. Anterior scalene m.

Figure 8.10 An anterolateral view of the neck muscles.

1. Thyroid cartilage
2. Thyrohyoid m.
3. Sternocleidomastoid m.
4. Levator scapulae m.
5. Omohyoid m. (superior belly)
6. Omohyoid m. (inferior belly)
7. Clavicle
8. Subclavius m.

74

Table 8.1 Muscles of facial expression.

Face Muscle	Origin(s)	Insertion(s)	Action	Innervation
Frontalis	Galea aponeurotica	Skin of eyebrow	Wrinkles forehead and elevates eyebrow	Facial n.
Orbicularis oculi	Bones of orbit	Tissue of eyelid	Closes eyes	Facial n.
Nasalis	Maxilla and nasal cartilage	Aponeurosis of nose	Compresses nostrils	Facial n.
Orbicularis oris	Maxilla and mandible	Lips	Closes and purses lips	Facial n.
Levator labii superioris	Upper maxilla	Orbicularis oris and skin above lips	Elevates upper lip	Facial n.
Zygomaticus	Zygomatic bone	Superior corner of orbicularis oris	Elevates corner of mouth	Facial n.
Depressor labii inferioris	Mandible	Inferior corner of orbicularis oris	Depresses corner of mouth	Facial n.
Mentalis	Mandible	Skin of lower lip	Elevates and protrudes lower lip	Facial n.
Buccinator	Maxilla and mandible	Orbicularis oris	Compresses cheek	Facial n.

Table 8.2 Muscles of mastication.

Chewing Muscle	Origin(s)	Insertion(s)	Action	Innervation
Temporalis	Temporal fossa	Coronoid process of mandible	Elevates mandible	Trigeminal n.
Masseter	Zygomatic arch	Lateral ramus of mandible	Elevates mandible	Trigeminal n.
Medial pterygoid	Sphenoid bone	Medial ramus of mandible	Elevates and laterally moves mandible	Trigeminal n.
Lateral pterygoid	Sphenoid bone	Anterior side of condylar process of mandible	Protracts mandible	Trigeminal n.

Table 8.3 Muscles of the neck.

Neck Muscle	Origin(s)	Insertion(s)	Action	Innervation
Sternocleidomastoid	Sternum and clavicle	Mastoid process of temporal bone	Flexes neck; rotates head to side	Accessory n.
Digastric	Inferior border of mandible and mastoid process of temporal bone	Hyoid bone	Depresses mandible; elevates hyoid bone	Trigeminal n. (ant. belly) Facial n. (post. belly)
Mylohyoid	Inferior border of mandible	Hyoid bone and median raphe	Elevates hyoid bone and floor of mouth	Trigeminal n.
Geniohyoid	Medial surface of mandible at chin	Hyoid bone	Elevates hyoid bone	Spinal n. (C1)
Stylohyoid	Styloid process of temporal bone	Hyoid bone	Elevates and retracts tongue	Facial n.
Sternohyoid	Manubrium	Hyoid bone	Depresses hyoid bone	Spinal n. (C1-C3)
Sternothyroid	Manubrium	Thyroid cartilage	Depresses thyroid cartilage	Spinal n. (C1-C3)
Thyrohyoid	Thyroid cartilage	Hyoid bone	Depresses hyoid bone; elevates thyroid cartilage	Spinal n. (C1)
Omohyoid	Superior border of scapula	Hyoid bone	Depresses hyoid bone	Spinal n. (C1-C3)

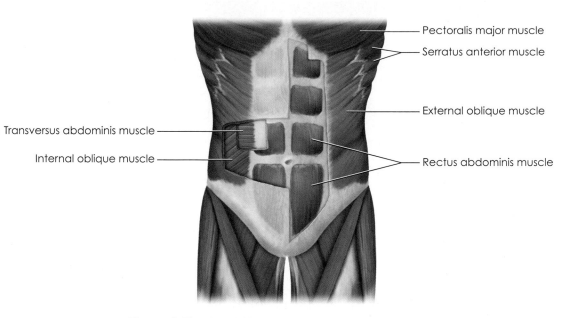

Figure 8.12 An anterior view of the muscles of the trunk.

Pectoralis major muscle

Serratus anterior muscle

External oblique muscle

Rectus abdominis muscle

Transversus abdominis muscle

Internal oblique muscle

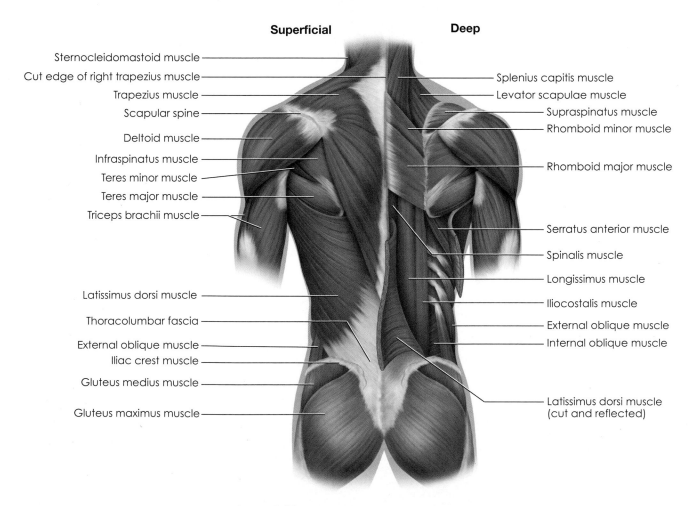

Superficial

Deep

Sternocleidomastoid muscle

Cut edge of right trapezius muscle

Trapezius muscle

Scapular spine

Deltoid muscle

Infraspinatus muscle

Teres minor muscle

Teres major muscle

Triceps brachii muscle

Latissimus dorsi muscle

Thoracolumbar fascia

External oblique muscle

Iliac crest muscle

Gluteus medius muscle

Gluteus maximus muscle

Splenius capitis muscle

Levator scapulae muscle

Supraspinatus muscle

Rhomboid minor muscle

Rhomboid major muscle

Serratus anterior muscle

Spinalis muscle

Longissimus muscle

Iliocostalis muscle

External oblique muscle

Internal oblique muscle

Latissimus dorsi muscle
(cut and reflected)

Figure 8.13 A posterior view of the muscles of the trunk.

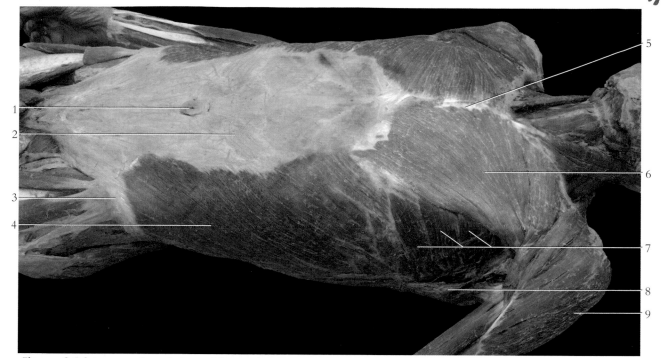

Figure 8.14 An anterolateral view of the trunk.

1. Umbilicus
2. Rectus sheath
3. Anterior superior iliac spine
4. External oblique m.
5. Sternum
6. Pectoralis major m.
7. Serratus anterior mm.
8. Latissimus dorsi m.
9. Deltoid m.

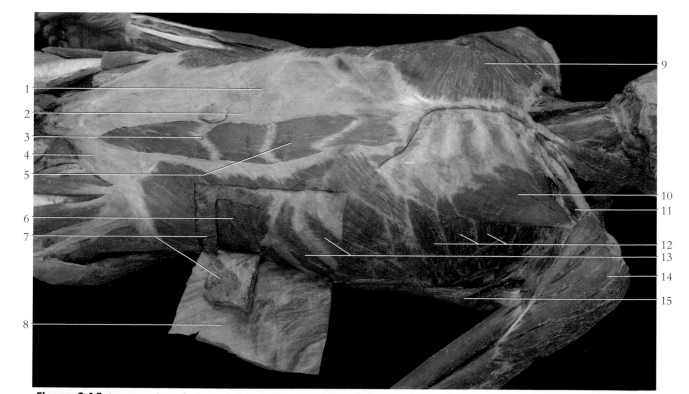

Figure 8.15 An anterolateral view of the trunk.

1. Rectus sheath
2. Linea alba
3. Tendinous intersection
4. Inguinal ligament
5. Rectus abdominis m.
6. Transverse abdominis m.
7. Internal oblique m.
8. External oblique m.
9. Pectoralis major m.
10. Pectoralis minor m.
11. Subclavius m.
12. Serratus anterior mm.
13. External intercostal mm.
14. Deltoid m.
15. Latissimus dorsi m.

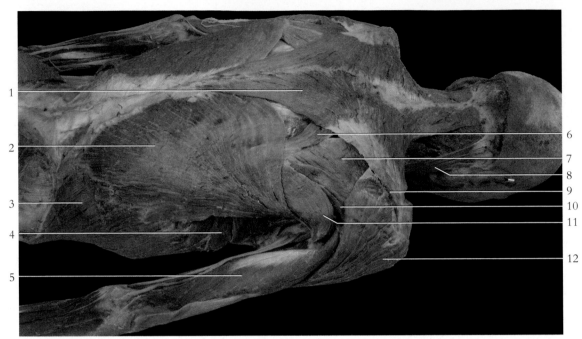

Figure 8.16 A posterolateral view of the trunk.

1. Trapezius m.
2. Latissimus dorsi m.
3. External oblique m.
4. Serratus anterior m.
5. Triceps brachii m.
6. Rhomboid major m.
7. Infraspinatus m.
8. Sternocleidomastoid m.
9. Spine of scapula
10. Teres minor m.
11. Teres major m.
12. Deltoid m.

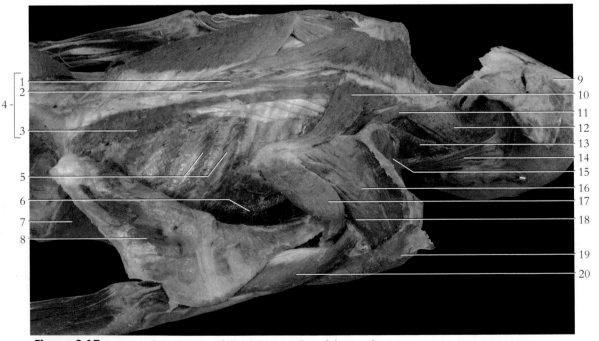

Figure 8.17 A posterolateral view of the deep muscles of the trunk.

1. Spinalis m. (cut)
2. Longissimus m.
3. Iliocostalis m.
4. Erecter spinae muscle group
5. External intercostal mm.
6. Serratus anterior m.
7. External oblique m.
8. Latissimus dorsi m. (cut and reflected)
9. Trapezius m. (cut and reflected)
10. Rhomboid major m.
11. Rhomboid minor m.
12. Splenius capitis m.
13. Levator scapulae m.
14. Sternocleidomastoid m.
15. Supraspinatus m.
16. Infraspinatus m.
17. Teres major m.
18. Teres minor m.
19. Deltoid m. (cut and reflected)
20. Triceps brachii m.

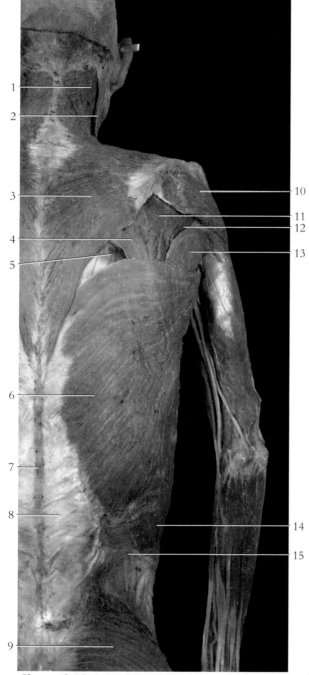

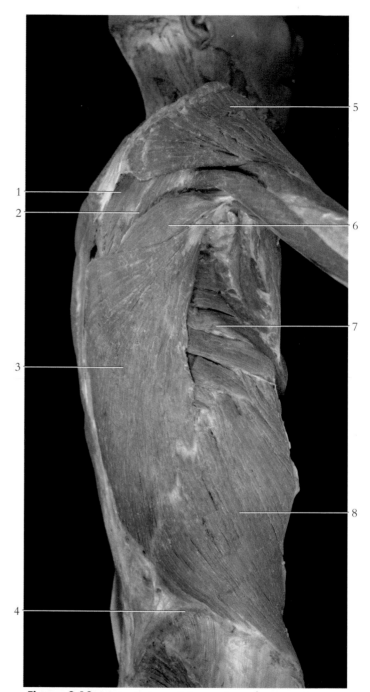

Figure 8.18 A posterior view of the trunk.

1. Splenius capitis m.
2. Sternocleidomastoid m.
3. Trapezius m.
4. Rhomboid major m.
5. Triangle of ausculation
6. Latissimus dorsi m.
7. Vertebral column
8. Thoracolumbar facia
9. Gluteus maximus m.
10. Deltoid m.
11. Infraspinatus m.
12. Teres minor m.
13. Teres major m.
14. External oblique m.
15. Iliac crest

Figure 8.19 A lateral view of the trunk.

1. Infraspinatus m.
2. Teres minor m.
3. Latissimus dorsi m.
4. Iliac crest
5. Deltoid m.
6. Teres major m.
7. Serratus anterior m.
8. External oblique m.

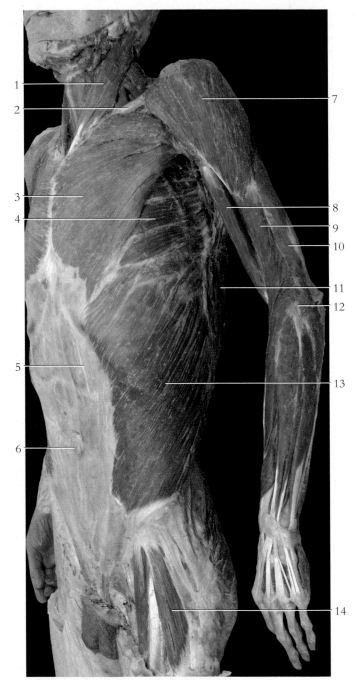

Figure 8.20 An anterolateral view of the trunk and arm.
1. Sternocleidomastoid m.
2. Clavicle
3. Pectoralis major m.
4. Serratus anterior m.
5. Rectus sheath
6. Umbilicus
7. Deltoid m.
8. Biceps brachii m.
9. Brachialis m.
10. Triceps brachii m.
11. Latissimus dorsi m.
12. Lateral epicondyle of humerus
13. External oblique m.
14. Tensor fascia latae m.

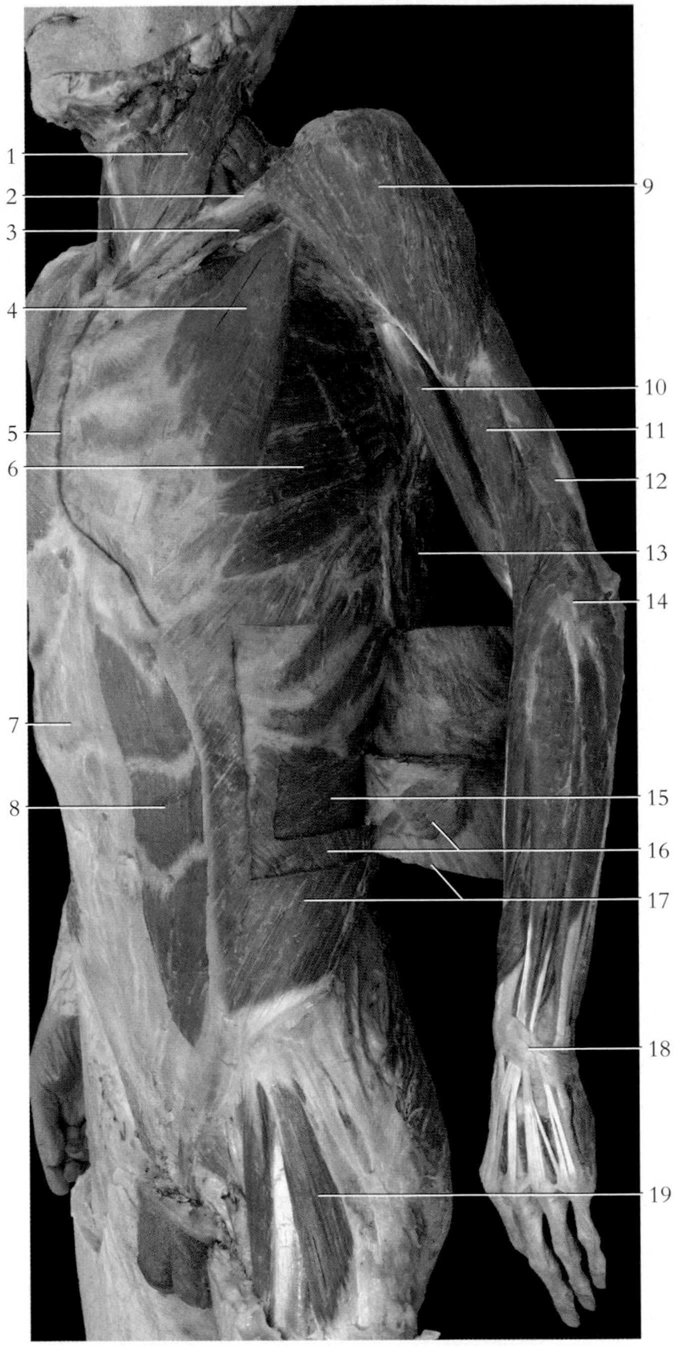

Figure 8.21 An anterolateral view of the trunk and arm.
1. Sternocleidomastoid m.
2. Clavicle
3. Subclavius m.
4. Pectoralis minor m.
5. Pectoralis major m. (cut)
6. Serratus anterior m.
7. Rectus sheath
8. Rectus abdominis m.
9. Deltoid m.
10. Biceps brachii m.
11. Brachialis m.
12. Triceps brachii m.
13. Latissimus dorsi m.
14. Lateral epicondyle of humerus
15. Transverse abdominis m.
16. Internal oblique m. (section cut)
17. External oblique m. (section cut)
18. Extensor retinaculum
19. Tensor fascia latae m.

Table 8.4 Muscles of the abdominal wall.

Abdominal Muscle	Origin(s)	Insertion(s)	Action	Innervation
External abdominal oblique	Lower eight ribs	Iliac crest and linea alba	Compresses abdomen; lateral rotation	Intercostal nn. Iliohypogastric n. Ilioinguinal n.
Internal abdominal oblique	Iliac crest, lumbodorsal fascia, inguinal ligament	Linea alba and costal cartilages	Compresses abdomen; lateral rotation	Intercostal nn. Iliohypogastric n. Ilioinguinal n.
Transversus abdominis	Iliac crest, lumbodorsal fascia, inguinal ligament, costal cartilages	Xiphoid process, linea alba, pubis	Compresses abdomen	Intercostal nn. Iliohypogastric n. Ilioinguinal n.
Rectus abdominis	Pubic crest and symphysis pubis	Xiphoid process and costal cartilages	Flexes vertebral column	Intercostal nn.

Table 8.5 Muscles that act on the pectoral girdle.

Pectoral Muscle	Origin(s)	Insertion(s)	Action	Innervation
Serratus anterior	Upper eight or nine ribs	Anterior medial border of scapula	Pulls scapula forward and upward	Long thoracic n.
Pectoralis minor	Sternal ends of third, fourth, and fifth ribs	Coracoid process of scapula	Pulls scapula forward and downward	Medial and lateral pectoral nn.
Subclavius	First rib	Subclavian groove of clavical	Depresses clavicle	Spinal nn. C5, C6
Trapezius	Occipital bone and spines of cervical and thoracic vertebrae	Clavicle, acromion and spine of scapula	Elevates, depresses, and adducts scapula; hyperextends neck; braces shoulder	Accessory n.
Levator scapulae	First to fourth cervical vertebrae	Superior border of scapula	Elevates scapula	Dorsal scapular n.
Rhomboid major	Spines of second to fifth thoracic vertebrae	Medial border of scapula	Elevates and adducts scapula	Dorsal scapular n.
Rhomboid minor	Seventh cervical and first thoracic vertebrae	Medial border of scapula	Elevates and adducts scapula	Dorsal scapular n.

Table 8.6 Muscles of the vertebral column.

Spinal Muscle	Origin(s)	Insertion(s)	Action	Innervation
Quadratus lumborum	Iliac crest and lower lumbar vertebrae	Twelfth rib and upper four lumbar vertebrae	Extends lumbar region; flexes vertebral column laterally	Intercostal n. T12 and Lumbar nn. L2-L4
Erector spinae				
Iliocostalis lumborum	Crest of ilium	Lower six ribs	Extends lumbar region	Post. rami of lumbar nn.
Iliocostalis thoracis	Lower six ribs	Upper six ribs	Extends thoracic region	Post. rami of thoracic nn.
Iliocostalis cervicis	Angles of third–sixth ribs	Transverse processes of fourth–sixth cervical vertebrae	Extends cervical region	Post. rami of cervical nn.

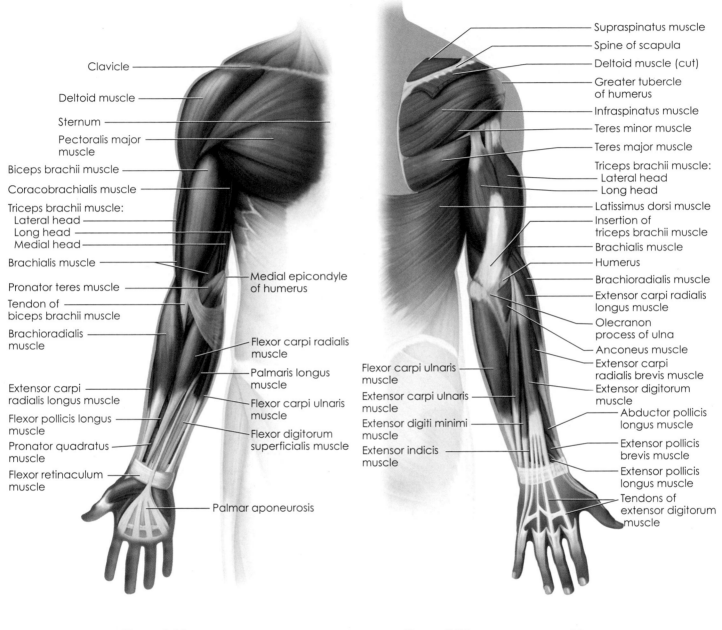

Clavicle

Deltoid muscle

Sternum

Pectoralis major muscle

Biceps brachii muscle

Coracobrachialis muscle

Triceps brachii muscle:
Lateral head
Long head
Medial head

Brachialis muscle

Pronator teres muscle

Tendon of biceps brachii muscle

Brachioradialis muscle

Extensor carpi radialis longus muscle

Flexor pollicis longus muscle

Pronator quadratus muscle

Flexor retinaculum muscle

Medial epicondyle of humerus

Flexor carpi radialis muscle

Palmaris longus muscle

Flexor carpi ulnaris muscle

Flexor digitorum superficialis muscle

Palmar aponeurosis

Supraspinatus muscle

Spine of scapula

Deltoid muscle (cut)

Greater tubercle of humerus

Infraspinatus muscle

Teres minor muscle

Teres major muscle

Triceps brachii muscle:
Lateral head
Long head

Latissimus dorsi muscle

Insertion of triceps brachii muscle

Brachialis muscle

Humerus

Brachioradialis muscle

Extensor carpi radialis longus muscle

Olecranon process of ulna

Anconeus muscle

Extensor carpi radialis brevis muscle

Extensor digitorum muscle

Abductor pollicis longus muscle

Extensor pollicis brevis muscle

Extensor pollicis longus muscle

Tendons of extensor digitorum muscle

Flexor carpi ulnaris muscle

Extensor carpi ulnaris muscle

Extensor digiti minimi muscle

Extensor indicis muscle

Figure 8.22 An anterior view of the muscles of the right upper arm.

Figure 8.23 A posterior view of the muscles of the right upper arm.

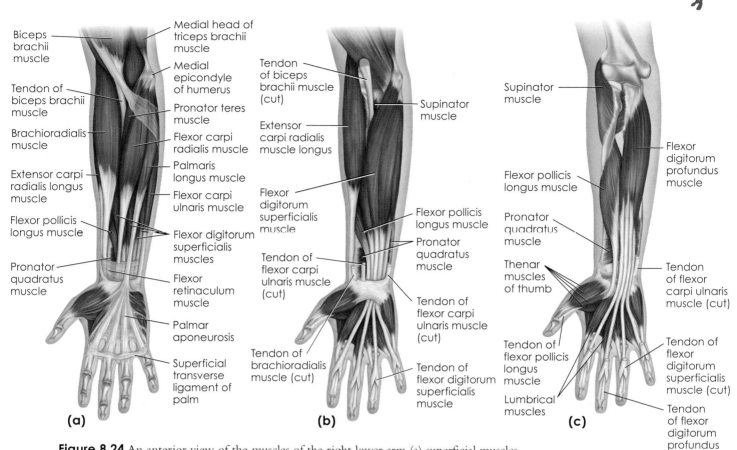

Figure 8.24 An anterior view of the muscles of the right lower arm (a) superficial muscles, (b) intermediate muscles, and (c) deep muscles.

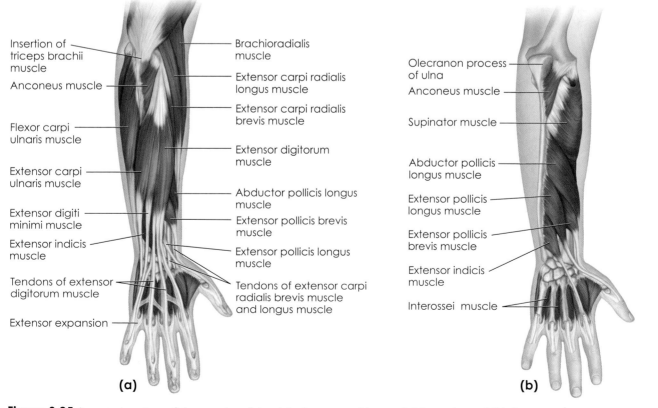

Figure 8.25 A posterior view of the muscles of the right lower arm (a) superficial muscles and (b) deep muscles.

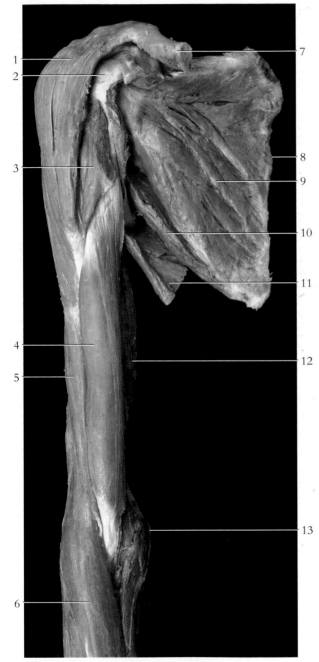

Figure 8.26 An anterior view of the right upper arm and shoulder (disarticulated).

1. Deltoid m.
2. Coracoid process of scapula
3. Pectoralis major m. (cut)
4. Biceps brachii m.
5. Brachialis m.
6. Brachioradialis m.
7. Clavicle (cut)
8. Medial margin of scapula
9. Subscapularis m.
10. Teres major m.
11. Latissimus dorsi m. (cut)
12. Triceps brachii m.
13. Medial epicondyle of humerus

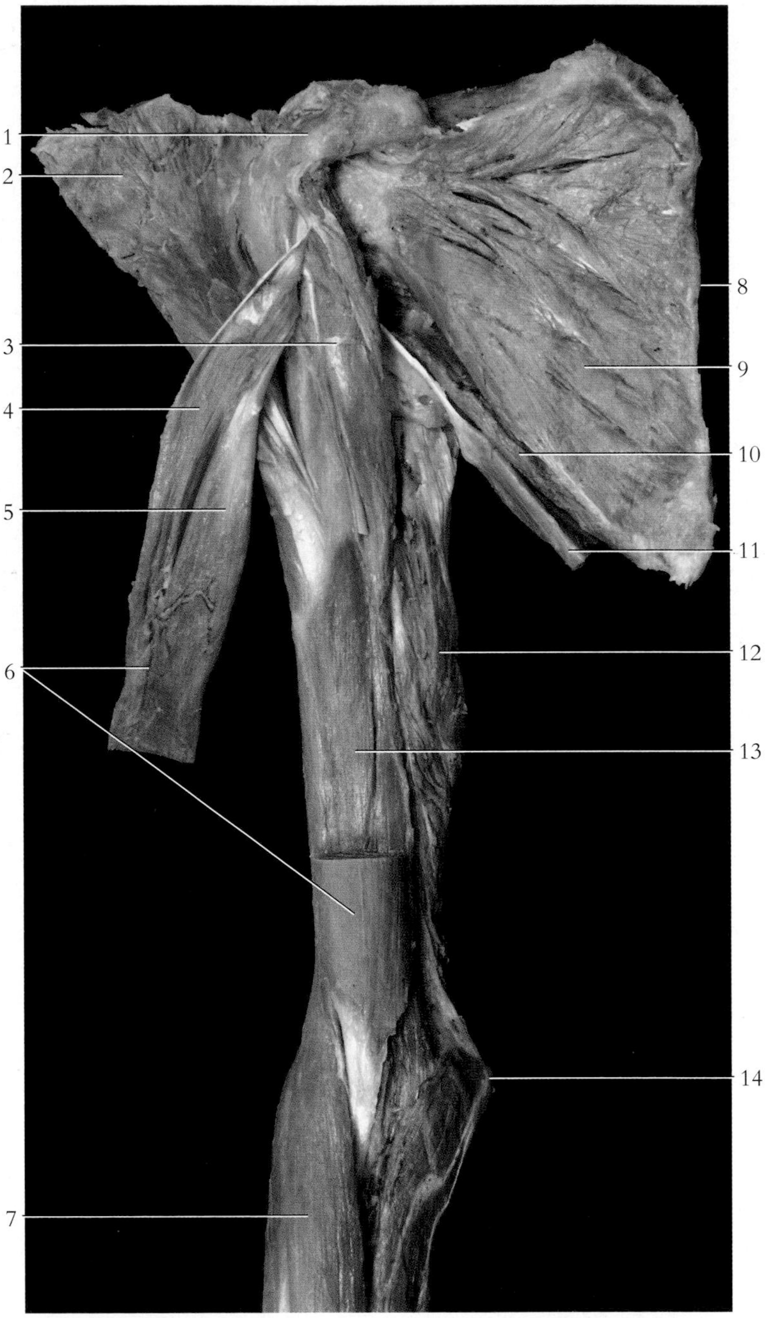

Figure 8.27 An anterior view of the right upper arm and shoulder (disarticulated).

1. Coracoid process of scapula
2. Deltoid m. (cut and reflected)
3. Coracobrachialis m.
4. Short head of biceps brachii m.
5. Long head of biceps brachii m.
6. Biceps brachii m. (cut and reflected)
7. Brachioradialis m.
8. Medial margin of scapula
9. Subscapularis m.
10. Teres major m.
11. Latissimus dorsi m. (cut)
12. Triceps brachii m.
13. Brachialis m.
14. Medial epicondyle of humerus

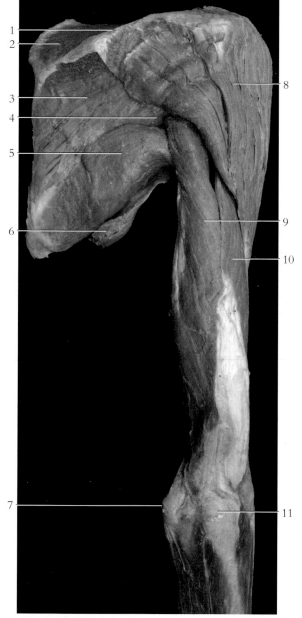

Figure 8.28 A posterior view of the right upper arm and shoulder (disarticulated).

1. Spine of scapula
2. Supraspinatus m.
3. Infraspinatus m.
4. Teres minor m.
5. Teres major m.
6. Latissimus dorsi m. (cut)
7. Medial epicondyle of humerus
8. Deltoid m.
9. Triceps brachii m. (long head)
10. Triceps brachii m. (lateral head)
11. Olecranon process of ulna

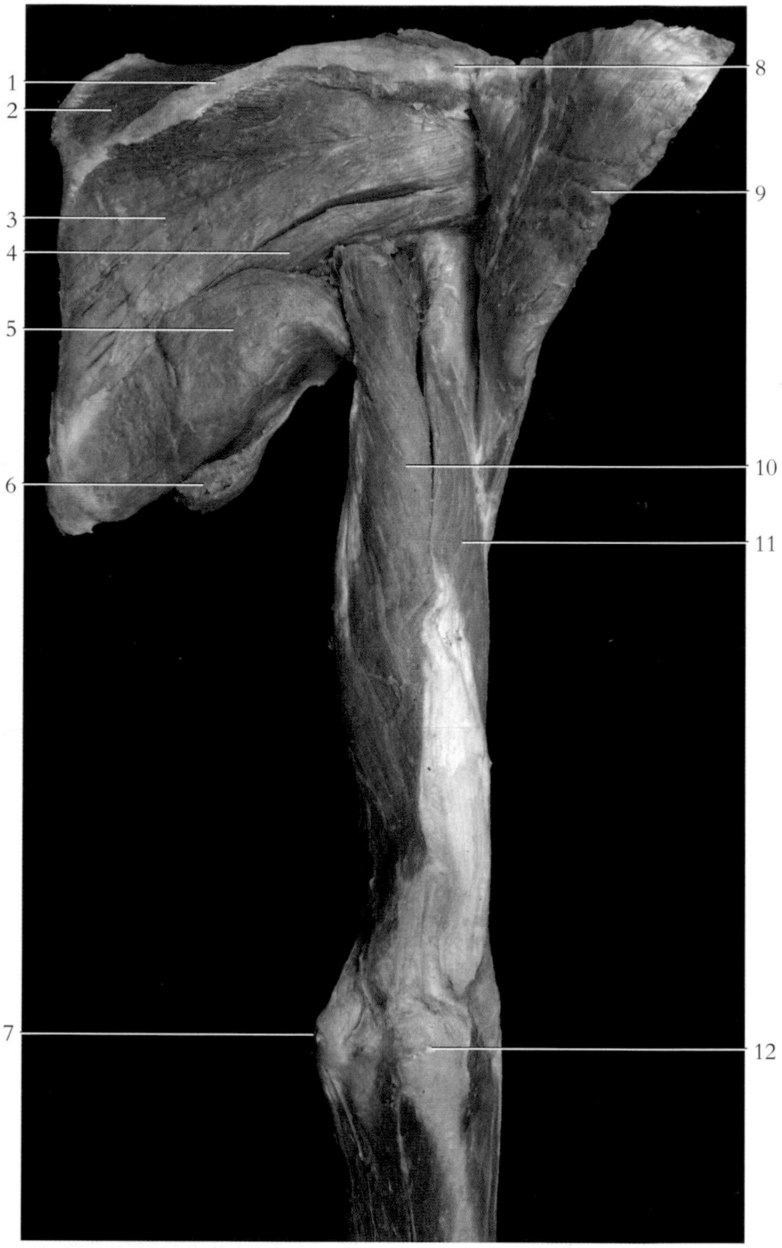

Figure 8.29 A posterior deep view of the right upper arm and shoulder (disarticulated).

1. Spine of scapula
2. Supraspinatus m.
3. Infraspinatus m.
4. Teres minor m.
5. Teres major m.
6. Latissimus dorsi m. (cut)
7. Medial epicondyle of humerus
8. Acromion
9. Deltoid m. (cut and reflected)
10. Triceps brachii m. (long head)
11. Triceps brachii m. (lateral head)
12. Olecranon process of ulna

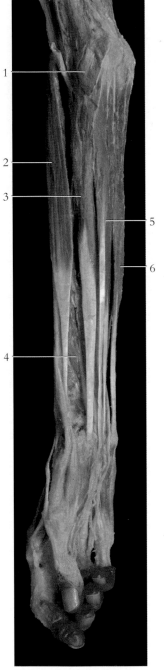

Figure 8.30 An anterior view of the superficial muscles of the right forearm.
1. Pronator teres m.
2. Brachioradialis m.
3. Flexor carpi radialis m.
4. Flexor pollicis longus m.
5. Palmaris longus m.
6. Flexor carpi ulnaris m.

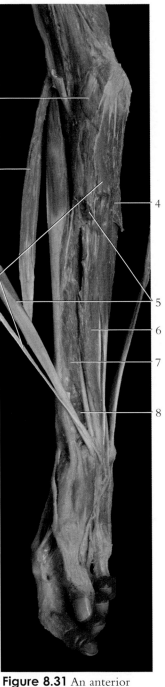

Figure 8.31 An anterior view of the muscles of the right forearm.
1. Pronator teres m.
2. Brachioradialis m. (cut and reflected)
3. Palmaris longus m. (cut and reflected)
4. Flexor carpi ulnaris m. (cut)
5. Flexor carpi radialis m. (cut and reflected)
6. Flexor digitorum superficialis m.
7. Flexor pollicis longus m.
8. Pronator quadratus m.

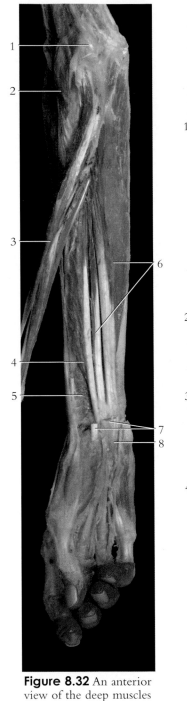

Figure 8.32 An anterior view of the deep muscles of the right forearm.
1. Medial epicondyle
2. Pronator teres m.
3. Flexor digitorum superficialis m. (cut and reflected)
4. Flexor pollicis longus m.
5. Pronator quadratus m.
6. Flexor digitorum profundus m.
7. Tendons of flexor carpi radialis and flexor digitorum superficialis mm. (cut)
8. Carpal tunnel

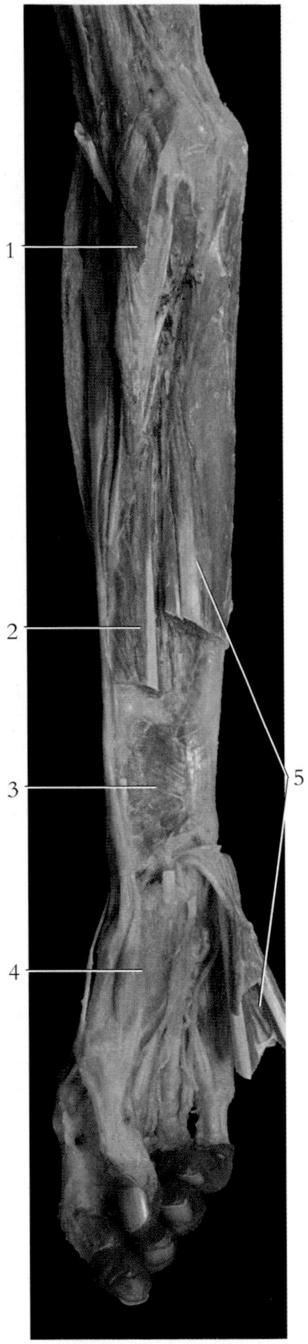

Figure 8.33 An anterior view of the deep muscles of the right forearm.
1. Pronator teres m.
2. Flexor pollicis longus m. (cut)
3. Pronator quadratus m.
4. Thenar mm.
5. Flexor digitorum superficialis and profundus mm. (cut and reflected)

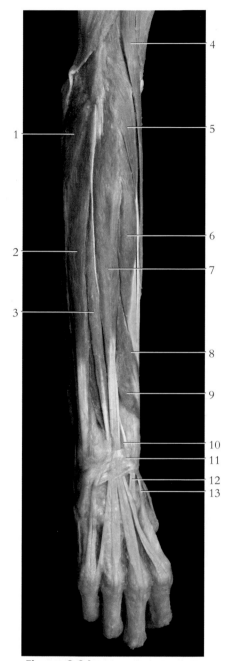

Figure 8.34 A posterior view of the superficial muscles of the right forearm.
1. Anconeus m.
2. Extensor carpi ulnaris m.
3. Extensor digiti minimi m.
4. Brachioradialis m.
5. Extensor carpi radialis longus m.
6. Extensor carpi radialis brevis m.
7. Extensor digitorum m.
8. Abductor pollicis longus m.
9. Extensor pollicis brevis m.
10. Extensor pollicis longus m.
11. Extensor retinaculum
12. Tendon of extensor carpi radialis brevis
13. Tendon of extensor carpi radialis longus

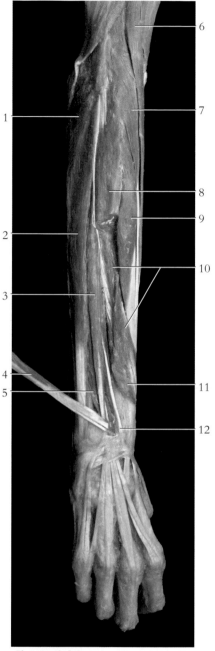

Figure 8.35 A posterior view of the muscles of the right forearm.
1. Anconeus m.
2. Extensor carpi ulnaris m.
3. Extensor digiti minimi m.
4. Extensor digitorum m. (cut and reflected)
5. Extensor indicis m.
6. Brachioradialis m.
7. Extensor carpi radialis longus m.
8. Extensor digitorum m. (cut)
9. Extensor carpi radialis brevis m.
10. Abductor pollicis longus m.
11. Extensor pollicis brevis m.
12. Tendon of extensor pollicis longus

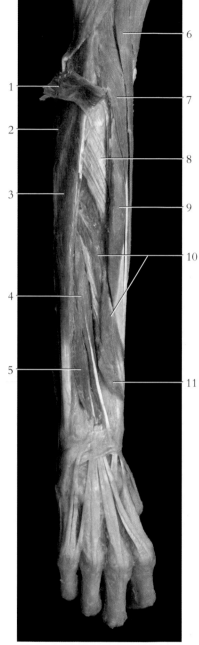

Figure 8.36 A posterior view of the deep muscles of the right forearm.
1. Extensor digitorum m. (cut and reflected)
2. Anconeus m.
3. Extensor carpi ulnaris m.
4. Extensor pollicis longus m.
5. Extensor indicis m.
6. Brachioradialis m.
7. Extensor carpi radialis longus m.
8. Supinator m.
9. Extensor carpi radialis brevis m.
10. Abductor pollicis longus m.
11. Extensor pollicis brevis m.

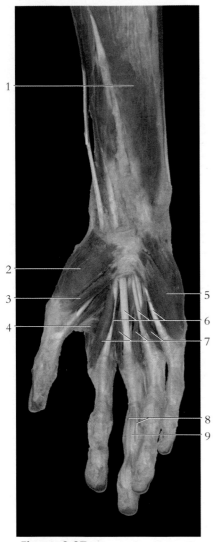

Figure 8.37 An anterior view of right hand.
1. Flexor carpi ulnaris m.
2. Abductor pollicis brevis m.
3. Flexor pollicis brevis m.
4. Adductor pollicis m.
5. Hypothenar mm.
6. Tendons of flexor digitorum superficialis
7. Lumbrical mm.
8. Flexor digitorum superficialis tendon (bifurcated for insertion)
9. Tendon of flexor digitorum profundus

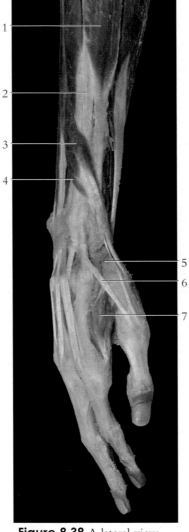

Figure 8.38 A lateral view of right hand.
1. Extensor carpi radialis longus m.
2. Tendon of extensor carpi radialis brevis m.
3. Abductor pollicis longus m.
4. Extensor pollicis brevis m.
5. Anatomical snuff box
6. Tendon of extensor pollicis longus m.
7. First dorsal interosseus m.

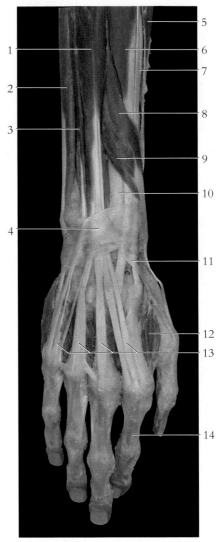

Figure 8.39 A posterior view of right hand.
1. Extensor digitorum m.
2. Extensor carpi ulnaris m.
3. Extensor digiti minimi m.
4. Extensor retinaculum
5. Brachioradialis m.
6. Extensor carpi radialis brevis m.
7. Tendon of extensor carpi radialis longus m.
8. Abductor pollicis longus m.
9. Extensor pollicis brevis m.
10. Radius
11. Tendon of extensor pollicis longus m.
12. First dorsal interosseus m.
13. Extensor digitorum tendons
14. Extensor expansion

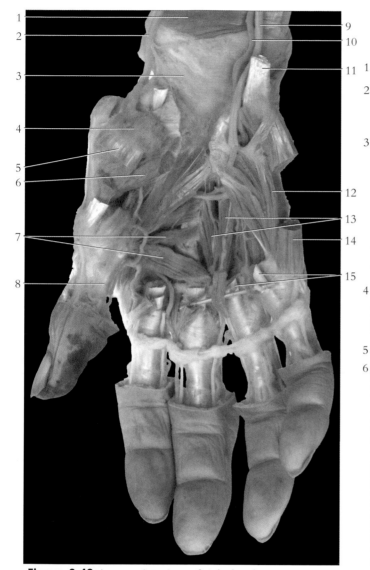

Figure 8.40 An anterior view of right hand.
1. Pronator quadratus m.
2. Radial artery
3. Flexor synovial sheath
4. Abductor pollicis brevis m.
5. Opponens pollicis m.
6. Flexor pollicis brevis m.
7. Adductor pollicis mm.
8. Tendon of flexor pollicis longus m.
9. Ulnar nerve
10. Ulnar artery
11. Tendon of flexor carpi ulnaris m.
12. Opponens digiti minimi m.
13. Lumbrical mm.
14. Abductor digiti minimi and flexor
    digiti minimi mm. (cut)
15. Tendons of deep digital flexor m. (cut)

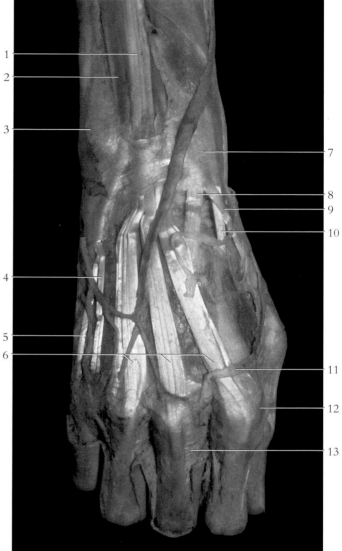

Figure 8.41 A posterior view of right hand showing
superficial structures.
1. Tendon of extensor digitorum m.
2. Extensor digiti minimi m.
3. Styloid process of ulna
4. Tendon of extensor digitorum m.
5. Tendon of extensor digiti minimi m.
6. Tendons of extensor digitorum mm.
7. Extensor retinaculum
8. Tendon of extensor carpi radialis brevis m.
9. Tendon of extensor pollicis longus m.
10. Tendon of extensor carpi radialis longus m.
11. Superficial venous arch
12. Articular capsule of metacarpophalangeal joint
13. Fibrous digital sheath

Table 8.7 Muscles of the shoulder and axilla.

Axial or Scapular Muscle	Origin(s)	Insertion(s)	Action	Innervation
Pectoralis major	Clavicle, sternum, costal cartilages of second to sixth ribs	Greater tubercle of humerus	Flexes, adducts, and rotates shoulder joint medially	Medial and lateral pectoral nn.
Latissimus dorsi	Spines of sacral, lumbar, and lower thoracic vertebrae; lower ribs	Intertubercular groove of humerus	Extends, adducts, and rotates shoulder joint medially; adducts shoulder joint	Thoracodorsal n.
Deltoid	Clavicle, acromion and spine of scapula	Deltoid tuberosity of humerus	Abducts, extends, or flexes shoulder joint	Axillary n.
Supraspinatus	Supraspinous fossa of scapula	Greater tubercle of humerus	Abducts shoulder joint	Suprascapular n.
Infraspinatus	Infraspinous fossa of scapula	Greater tubercle of humerus	Rotates shoulder joint laterally	Suprascapular n.
Teres major	Inferior angle and lateral border of scapula	Intertubercular groove of humerus	Extends, and rotates shoulder joint medially	Lower subscapular n.
Teres minor	Lateral border of scapula	Greater tubercle of humerus	Rotates shoulder joint laterally	Axillary n.
Subscapularis	Subscapular fossa	Lesser tubercle of humerus	Rotates shoulder joint medially	Subscapular n.
Coracobrachialis	Coracoid process of scapula	Body of humerus	Flexes and adducts shoulder joint	Musculocutaneous n.

Table 8.8 Muscles of the brachium (arm).

Brachial Muscle	Origin(s)	Insertion(s)	Action	Innervation
Biceps brachii	Coracoid process and tuberosity above glenoid fossa	Radial tuberosity	Flexes elbow joint; supinates forearm and hand at radioulnar joint	Musculocutaneous n.
Brachialis	Anterior body of humerus	Tuberosity of ulna	Flexes elbow joint	Musculocutaneous n.
Brachioradialis	Lateral supracondylar ridge of humerus	Proximal to styloid process of radius	Flexes elbow joint	Radial n.
Triceps brachii	Tuberosity below glenoid fossa: lateral and medial surfaces of humerus	Olecranon of ulna	Extends elbow joint	Radial n.
Anconeus	Lateral epicondyle of humerus	Olecranon of ulna	Extends elbow joint	Radial n.

Table 8.9 Muscles of the antebrachium (forearm).

Antebrachial Muscle	Origin(s)	Insertion(s)	Action	Innervation
Supinator	Lateral epicondyle of humerus and crest of ulna	Lateral surface of radius	Supinates forearm and hand	Radial n.
Pronator teres	Medial epicondyle of humerus	Lateral surface of radius	Pronates forearm and hand	Median n.
Pronator quadratus	Distal fourth of ulna	Distal fourth of radius	Pronates forearm and hand	Median n.
Flexor carpi radialis	Medial epicondyle of humerus	Base of second and third metacarpal bones	Flexes and abducts wrist	Median n.
Palmaris longus	Medial epicondyle of humerus	Palmar aponeurosis	Flexes wrist	Median n.
Flexor carpi ulnaris	Medial epicondyle of humerus and olecranon of ulna	Carpal and metacarpal bones	Flexes and adducts wrist	Ulnar n.
Flexor digitorum superficialis	Medial epicondyle of humerus and coronoid process of ulna	Middle phalanges of digits II–V	Flexes wrist and digits	Median n.
Flexor digitorum profundus	Proximal two-thirds of ulna and interosseous ligament	Distal phalanges of digits II–V	Flexes wrist and digits	Median and ulnar n.
Flexor pollicis longus	Body of radius and coronoid process of ulna	Distal phalanx of thumb	Flexes joints of thumb	Median n.
Extensor carpi radialis longus	Lateral supracondylar ridge of humerus	Second metacarpal bone	Extends and abducts wrist	Radial n.
Extensor carpi radialis brevis	Lateral epicondyle of humerus	Third metacarpal bone	Extends and abducts wrist	Radial n.
Extensor digitorum	Lateral epicondyle of humerus	Posterior surfaces of digits II–V	Extends wrist and phalanges	Radial n.
Extensor digiti minimi	Lateral epicondyle of humerus	Extensor aponeurosis of fifth digit	Extends joints of fifth digit and wrist	Radial n.
Extensor carpi ulnaris	Lateral epicondyle of humerus and olecranon of ulna	Base of fifth metacarpal bone	Extends and adducts wrist	Radial n.
Extensor pollicis longus	Lateral and posterior surface of ulna	Base of distal phalanx of thumb	Extends joints of thumb; abducts joints of hand	Radial n.
Extensor pollicis brevis	Distal body of radius and interosseous ligament	Base of proximal phalanx of thumb	Extends joints of thumb; abducts joints of hand	Radial n.
Abductor pollicis longus	Distal radius and ulna and interosseous ligament	Base of first metacarpal bone	Abducts joints of thumb and joints of hand	Radial n.

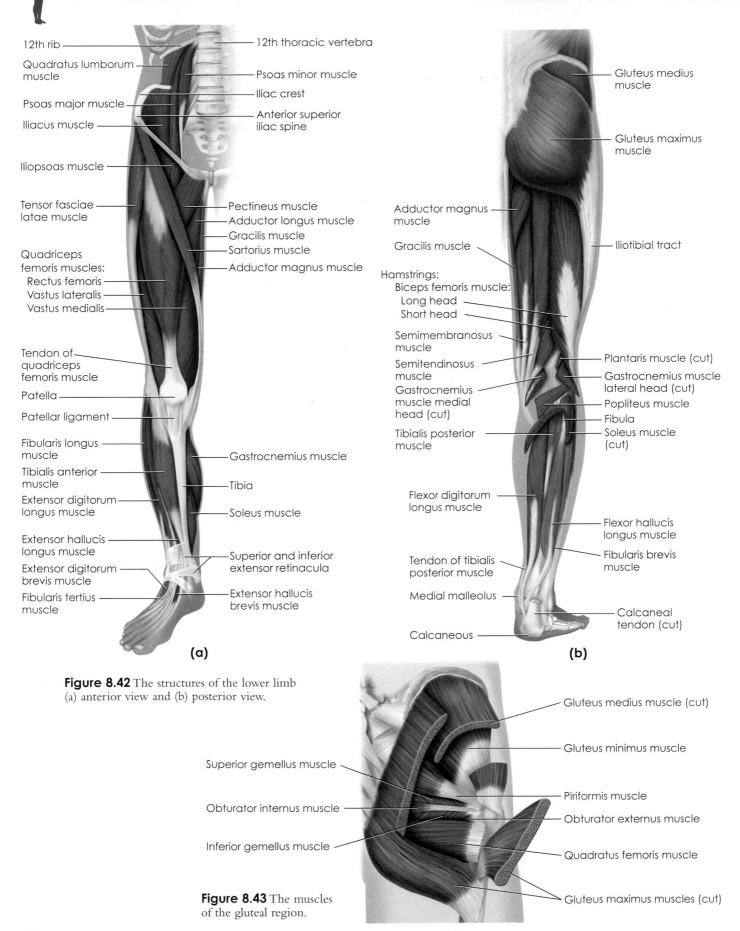

Figure 8.42 The structures of the lower limb (a) anterior view and (b) posterior view.

Figure 8.43 The muscles of the gluteal region.

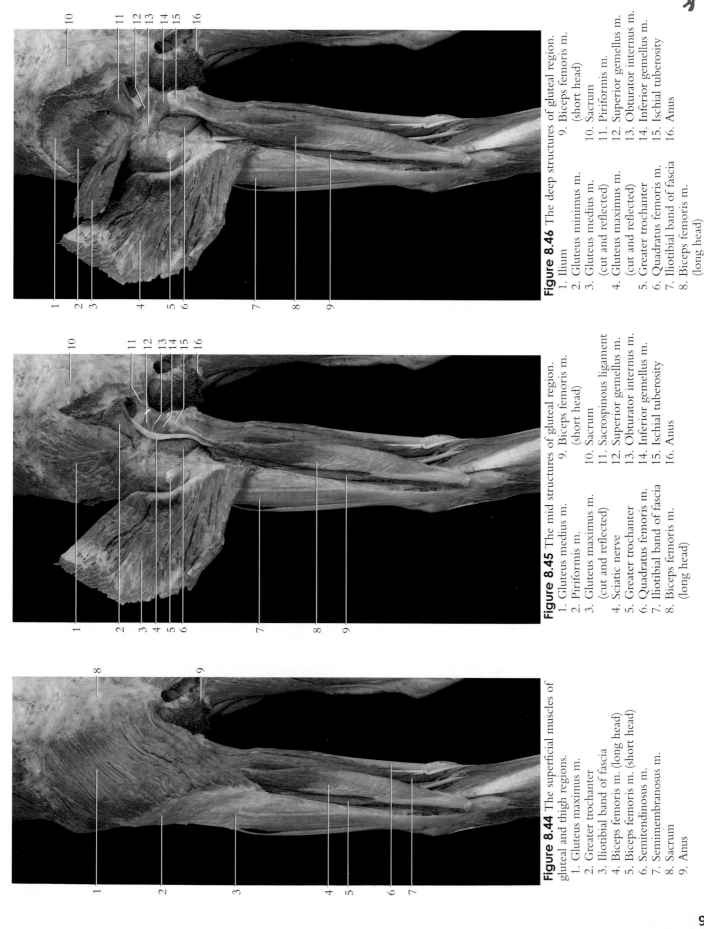

Figure 8.46 The deep structures of gluteal region.

1. Ilium
2. Gluteus minimus m.
3. Gluteus medius m.
   (cut and reflected)
4. Gluteus maximus m.
   (cut and reflected)
5. Greater trochanter
6. Quadratus femoris m.
7. Iliotibial band of fascia
8. Biceps femoris m.
   (long head)
9. Biceps femoris m.
   (short head)
10. Sacrum
11. Piriformis m.
12. Superior gemellus m.
13. Obturator internus m.
14. Inferior gemellus m.
15. Ischial tuberosity
16. Anus

Figure 8.45 The mid structures of gluteal region.

1. Gluteus medius m.
2. Piriformis m.
3. Gluteus maximus m.
   (cut and reflected)
4. Sciatic nerve
5. Greater trochanter
6. Quadratus femoris m.
7. Iliotibial band of fascia
8. Biceps femoris m.
   (long head)
9. Biceps femoris m.
   (short head)
10. Sacrum
11. Sacrospinous ligament
12. Superior gemellus m.
13. Obturator internus m.
14. Inferior gemellus m.
15. Ischial tuberosity
16. Anus

Figure 8.44 The superficial muscles of gluteal and thigh regions.

1. Gluteus maximus m.
2. Greater trochanter
3. Iliotibial band of fascia
4. Biceps femoris m. (long head)
5. Biceps femoris m. (short head)
6. Semitendinosus m.
7. Semimembranosus m.
8. Sacrum
9. Anus

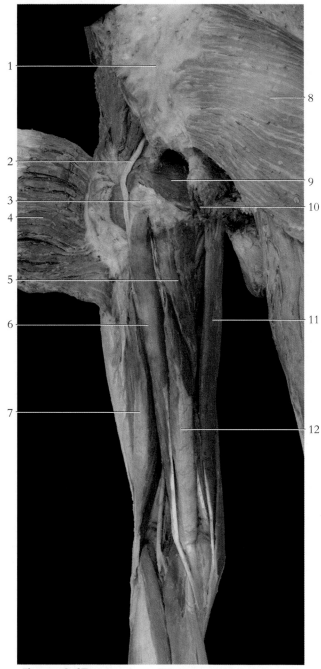

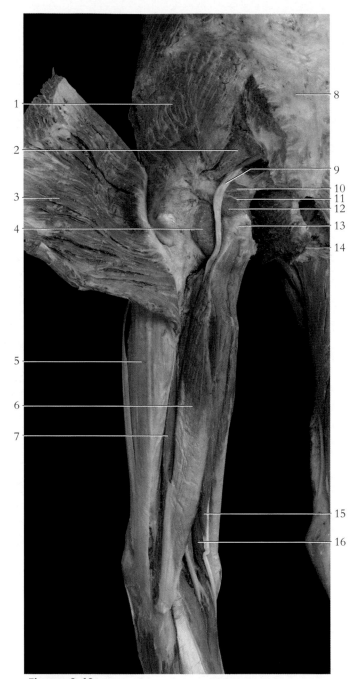

Figure 8.47 A posteromedial view of the posterior structures of thigh region.
1. Sacrum
2. Sciatic nerve
3. Ischial tuberosity
4. Gluteus maximus m. (left side) (cut and reflected)
5. Adductor magnus m.
6. Semitendinosus m.
7. Biceps femoris m.
8. Gluteus maximus m. (right side)
9. Obturator internus m.
10. Anus
11. Gracilis m.
12. Semimembranosus m.

Figure 8.48 A posterolateral view of the posterior structures of thigh region.
1. Gluteus medius m.
2. Piriformis m.
3. Gluteus maximus m. (left side) (cut and reflected)
4. Quadratus femoris m.
5. Iliotibial band
6. Biceps femoris m. (long head)
7. Biceps femoris m. (short head)
8. Sacrum
9. Sciatic nerve
10. Superior gemellus m.
11. Obturator internus m.
12. Inferior gemellus m.
13. Ischial tuberosity
14. Anus
15. Semitendinosus m.
16. Semimembranosus

Figure 8.49 An anterior view of the right femoral triangle.
1. Inguinal ligament
2. Lateral femoral cutaneous nerve
3. Superficial circumflex iliac artery
4. Iliopsoas m.
5. Femoral nerve
6. Femoral artery
7. Tensor fasciae latae m.
8. Sartorius m.
9. Rectus femoris m.
10. Femoral ring
11. Femoral vein
12. Pectineus m.
13. Great saphenous vein
14. Adductor longus m.

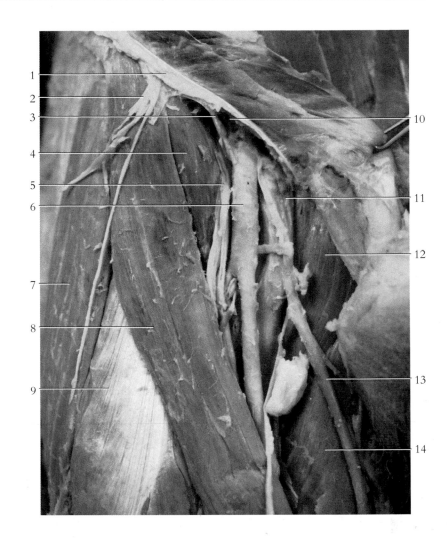

Table 8.10 Muscles of the hip.

Pelvic Muscle	Origin(s)	Insertion(s)	Action	Innervation
Iliacus	Iliac fossa	Lesser trochanter of femur, along with psoas major	Flexes at the hip joint; flexes joints of vertebral column	Femoral n.
Psoas major	Transverse process of lumbar vertebrae	Lesser trochanter of femur, along with iliacus	Flexes at the hip joint; flexes joints of vertebral column	Spinal nn. L2, L3
Gluteus maximus	Iliac crest, sacrum, coccyx, aponeurosis of lumbar region	Gluteal tuberosity and iliotibial tract	Extends and rotates thigh laterally at the hip joint	Inferior gluteal n.
Gluteus medius	Lateral surface of ilium	Greater trochanter of femur	Abducts and rotates thigh medially at the hip joint	Superior gluteal n.
Gluteus minimus	Lateral surface of lower half of ilium	Greater trochanter of femur	Abducts and rotates thigh medially at the hip joint	Superior gluteal n.
Tensor fasciae latae	Anterior border of ilium and iliac crest	Iliotibial tract	Abducts and rotates thigh medially at the hip joint	Superior gluteal n.

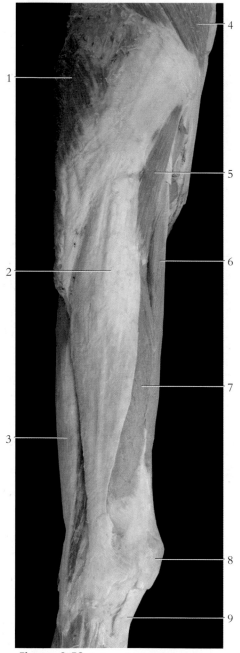

Figure 8.50 A lateral view of the right thigh.
1. Gluteus maximus m.
2. Iliotibial band
3. Biceps femoris m.
4. External oblique m.
5. Tensor fascia lata m.
6. Rectus femoris m.
7. Vastus lateralis m.
8. Patella
9. Patellar ligament

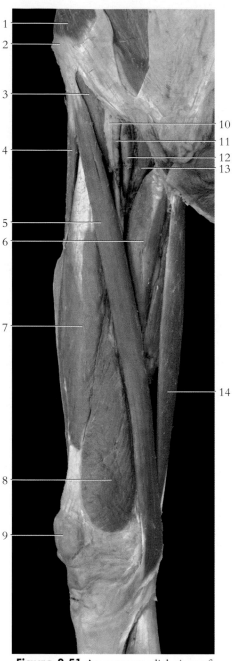

Figure 8.51 An anteromedial view of the right thigh.
1. External oblique m.
2. Anterior superior iliac spine
3. Iliopsoas m.
4. Tensor fascia lata m.
5. Sartorius m.
6. Adductor longus m.
7. Rectus femoris m.
8. Vastus medialis m.
9. Patella
10. Femoral nerve
11. Femoral artery
12. Femoral vein
13. Pectineus m.
14. Gracilis m.

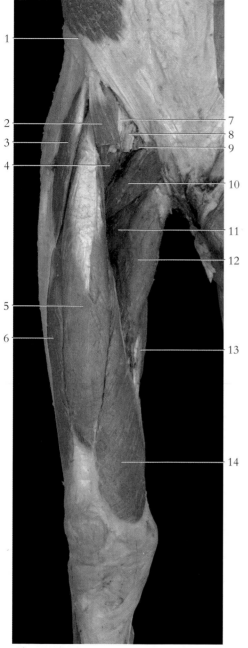

Figure 8.52 An anterior view of the right thigh.
1. Anterior superior iliac spine
2. Sartorius m. (cut)
3. Tensor fascia lata m.
4. Iliopsoas m.
5. Rectus femoris m.
6. Vastus lateralis m.
7. Femoral nerve (cut)
8. Femoral artery (cut)
9. Femoral vein (cut)
10. Pectineus m.
11. Adductor brevis m.
12. Adductor longus m.
13. Adductor magnus m.
14. Vastus medialis m.

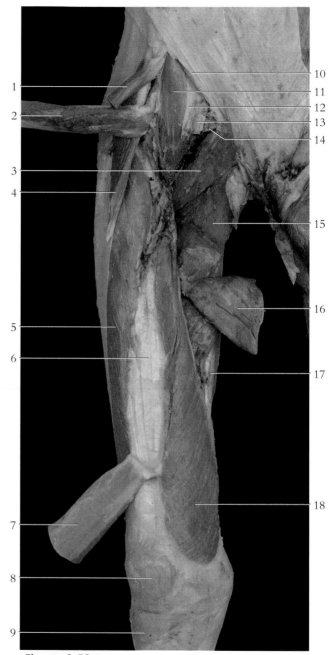

Figure 8.53 An anterior view of the deep structures of the right thigh.
1. Sartorius m. (cut and reflected)
2. Rectus femoris m. (cut and reflected)
3. Pectineus m.
4. Tensor fascia lata m.
5. Vastus lateralis m.
6. Vastus intermedius m.
7. Rectus femoris m. (cut and reflected)
8. Patella
9. Patellar ligament
10. Inguinal ligament
11. Iliopsoas m.
12. Femoral nerve (cut)
13. Femoral artery (cut)
14. Femoral vein (cut)
15. Adductor brevis m.
16. Adductor longus m. (cut and reflected)
17. Adductor magnus m.
18. Vastus medialis m.

Table 8.11 Medial muscles of the hip.

Adductor Muscle	Origin(s)	Insertion(s)	Action	Innervation
Gracilis	Inferior edge of symphysis pubis	Proximomedial surface of tibia	Adducts thigh at hip joint; flexes leg at knee joint	Obturator n.
Pectineus	Pectineal line of pubis	Distal to lesser trochanter of femur	Adducts and flexes thigh at hip joint	Femoral n.
Adductor longus	Pubis–below pubic crest	Linea aspera of femur	Adducts thigh at hip joint	Obturator n.
Adductor brevis	Inferior ramus of pubis	Linea aspera of femur	Adducts thigh at hip joint	Obturator n.
Adductor magnus	Inferior ramus of ischium and inferior ramus of pubis	Linea aspera and medial epicondyle of femur	Adducts thigh at hip joint	Obturator and tibial nn.

Table 8.12 Muscles of the knee.

Thigh Muscle	Origin(s)	Insertion(s)	Action	Innervation
Sartorius	Anterior superior iliac spine	Medial surface of tibia	Flexes leg and thigh; abducts rotates thigh laterally	Femoral n.
Quadriceps femoris		Patella by common tendon, which continues as patellar ligament to tibial tuberosity	Extends leg at knee joint	Femoral n.
Rectus femoris	Anterior inferior iliac spine			
Vastus lateralis	Inferior to greater trochanter and linea aspera of femur			
Vastus medialis	Medial surface and linea aspera of femur			
Vastus intermedius	Anterior and lateral surfaces of femur			
Biceps femoris	Long head—ischial tuberosity; short head—linea aspera of femur	Head of fibula and proximolateral part of tibia	Extends thigh at hip joint; flexes leg at knee joint	Tibial n.
Semitendinosus	Ischial tuberosity	Proximomedial surface of tibia	Extends thigh at hip joint; flexes leg at knee joint	Tibial n.
Semimembranosus	Ischial tuberosity	Proximomedial surface of tibia	Extends thigh at hip joint; flexes leg at knee joint	Tibial n.

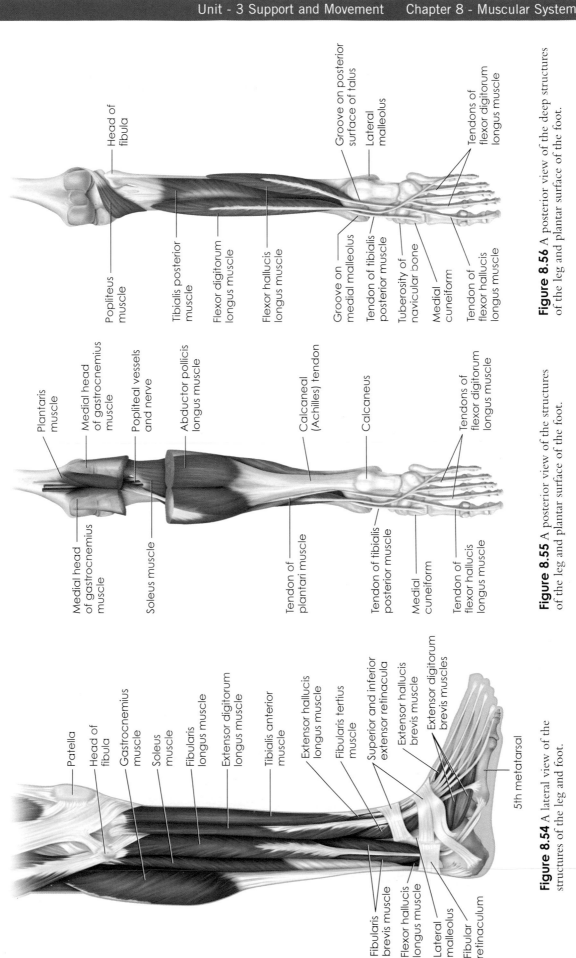

Head of fibula

Popliteus muscle

Tibialis posterior muscle

Flexor digitorum longus muscle

Flexor hallucis longus muscle

Groove on posterior surface of talus

Lateral malleolus

Tendons of flexor digitorum longus muscle

Groove on medial malleolus

Tendon of tibialis posterior muscle

Tuberosity of navicular bone

Medial cuneiform

Tendon of flexor hallucis longus muscle

Figure 8.56 A posterior view of the deep structures of the leg and plantar surface of the foot.

Plantaris muscle

Medial head of gastrocnemius muscle

Popliteal vessels and nerve

Abductor pollicis longus muscle

Calcaneal (Achilles) tendon

Calcaneus

Tendons of flexor digitorum longus muscle

Medial head of gastrocnemius muscle

Soleus muscle

Tendon of plantari muscle

Tendon of tibialis posterior muscle

Medial cuneiform

Tendon of flexor hallucis longus muscle

Figure 8.55 A posterior view of the structures of the leg and plantar surface of the foot.

Patella

Head of fibula

Gastrocnemius muscle

Soleus muscle

Fibularis longus muscle

Extensor digitorum longus muscle

Tibialis anterior muscle

Extensor hallucis longus muscle

Fibularis tertius muscle

Superior and inferior extensor retinacula

Extensor hallucis brevis muscle

Extensor digitorum brevis muscles

5th metatarsal

Fibularis brevis muscle

Flexor hallucis longus muscle

Lateral malleolus

Fibular retinaculum

Figure 8.54 A lateral view of the structures of the leg and foot.

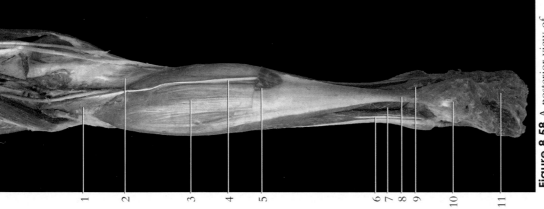

Figure 8.59 A posterior deep view of lower leg.

1. Soleus m. (cut)
2. Popliteus m.
3. Fibula
4. Tibial nerve
5. Flexor hallucis longus m.
6. Flexor digitorum longus m.
7. Fibularis longus m.
8. Calcaneus
9. Flexor digitorum brevis m.

Figure 8.58 A posterior view of lower leg.

1. Plantaris m.
2. Popliteus m.
3. Soleus m.
4. Plantaris tendon
5. Gastrocnemius m. (cut)
6. Fibularis longus and brevis mm.
7. Flexor hallucis longus m.
8. Tendo calcaneus (Achilles tendon)
9. Tendon of flexor hallucis longus m.
10. Calcaneus
11. Plantar aponeurosis

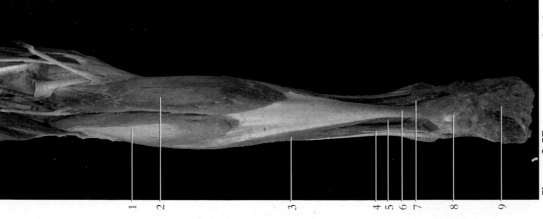

Figure 8.57 A posterior superficial view of lower leg.

1. Lateral head of gastrocnemius m.
2. Medial head of gastrocnemius m.
3. Soleus m.
4. Fibularis longus and brevis mm.
5. Flexor hallucis longus m.
6. Tendo calcaneus (Achilles tendon)
7. Tendon of flexor hallucis longus m.
8. Calcaneus
9. Plantar aponeurosis

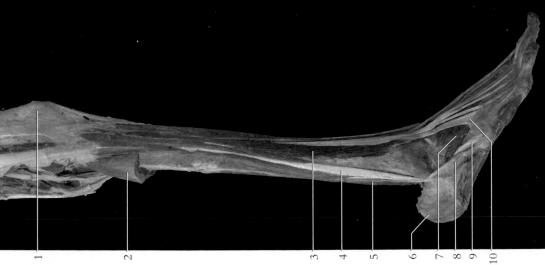

Figure 8.60 A lateral view of lower leg.
1. Patella
2. Soleus m. (cut)
3. Extensor digitorum longus m.
4. Tendon of fibularis longus m.
5. Fibularis brevis m.
6. Calcaneus
7. Extensor digitorum brevis m.
8. Tendon of fibularis longus m.
9. Tendon of fibularis brevis m.
10. Tendon of fibularis tertius m.

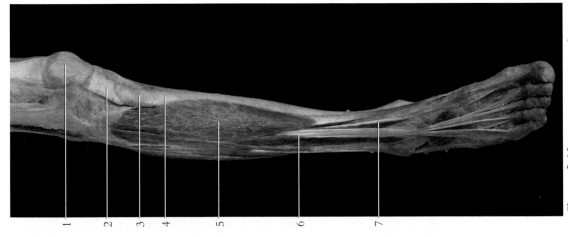

Figure 8.61 An anterior view of lower leg.
1. Patella
2. Patellar ligament
3. Tibial tuberosity
4. Tibia
5. Tibialis anterior m.
6. Tendon of extensor digitorum longus m.
7. Tendon of extensor hallucis longus m.

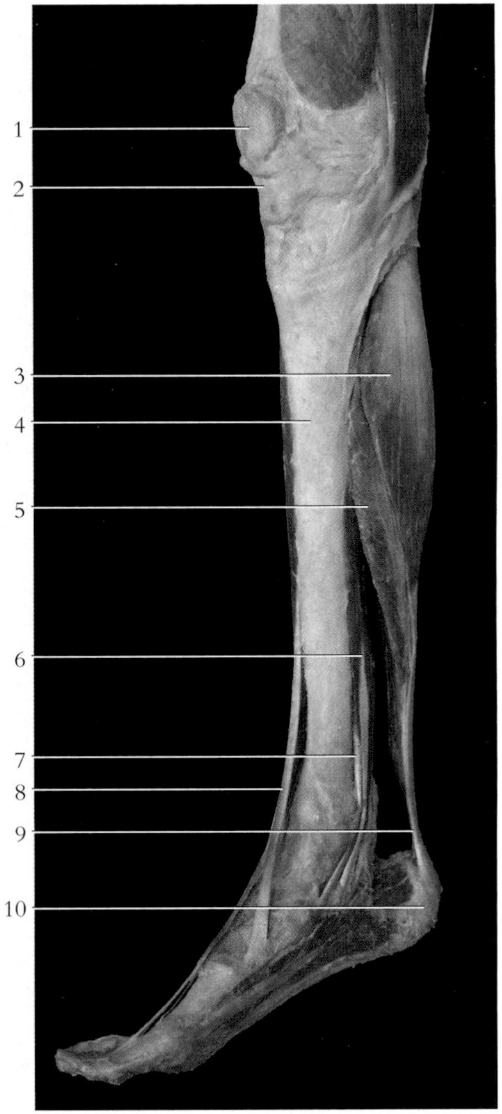

Figure 8.62 A medial view of lower leg.
1. Patella
2. Patellar ligament
3. Gastrocnemius m.
4. Tibia
5. Soleus m.
6. Flexor digitorum longus m.
7. Tendon of tibialis posterior m.
8. Tendon of tibialis anterior m.
9. Calcaneal tendon (Achilles tendon)
10. Calcaneus

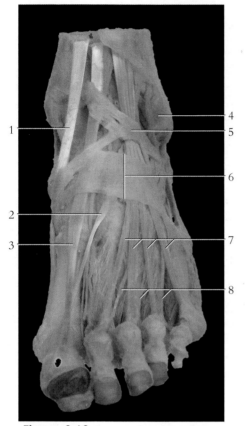

Figure 8.63 An anterior view of dorsum of foot.
1. Tendon of tibialis anterior m.
2. Tendon of extensor hallucis brevis m.
3. Tendon of extensor hallucis longus m.
4. Lateral malleolus
5. Superior extensor retinaculum
6. Inferior extensor retinaculum
7. Tendon of extensor digitorum longus m.
8. Tendon of extensor digitorum brevis m.

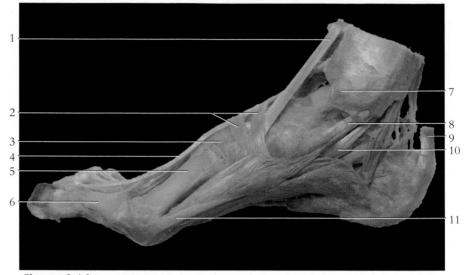

Figure 8.64 A medial view of the right foot.
1. Tendon of tibialis anterior m.
2. Extensor retinaculum
3. Medial cuneiform
4. Tendon of extensor hallucis longus m.
5. First metatarsal bone
6. Proximal phalanx of hallux
7. Medial malleolus of tibia
8. Tendon of tibialis posterior m.
9. Tendo calcaneus
10. Tendon of flexor digitorum longus m.
11. Abductor hallucis m.

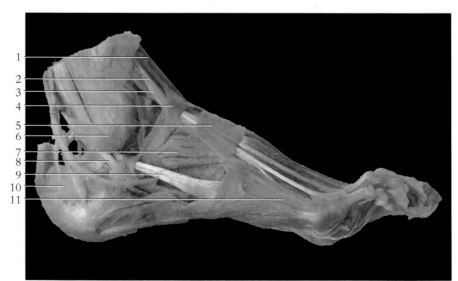

Figure 8.65 A lateral view of the right foot.
1. Tendon of tibialis anterior
2. Tendon of extensor digitorum longus m.
3. Tendon of fibularis tertius m.
4. Superior extensor retinaculum
5. Inferior extensor retinaculum
6. Lateral malleolus of fibula
7. Extensor digitorum brevis m.
8. Tendon of fibularis longus m.
9. Tendon of fibularis brevis m.
10. Calcaneus
11. Fifth metatarsal bone

Figure 8.66 A superficial
view of plantar region.
1. Flexor digiti
   minimi m.
2. Abductor digiti
   minimi m.
3. Plantar aponeurosis
4. Flexor hallucis brevis m.
5. Adductor hallucis m.
6. Calcaneus

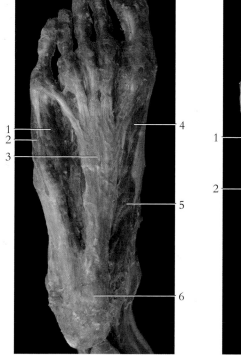

Figure 8.67 A view of
plantar region.
1. Flexor digiti minimi
   brevis m.
2. Abductor digiti
   minimi m.
3. Tendons of flexor
   digitorum brevis m.
4. Flexor digitorum
   brevis m.
5. Abductor hallucis m.
6. Plantar aponeurosis (cut)
7. Calcaneal tuberosity

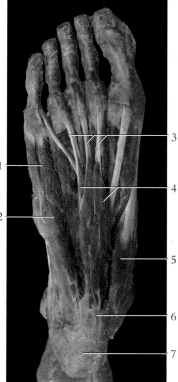

Figure 8.68 A deep
view of plantar region.
1. Interosseous m.
2. Quadratus plantae m.
3. Flexor digitorum
   brevis m. (cut and
   reflected)
4. Tendon of flexor
   hallucis longus m.
5. Oblique head of
   adductor hallucis m.
6. Lumbrical mm.
7. Flexor hallucis brevis m.
8. Tendons of flexor
   digitorum longus m.
9. Abductor hallucis m.

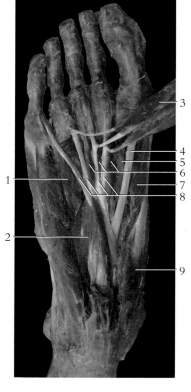

Figure 8.69 A deep
view of plantar region.
1. Interosseous mm.
2. Oblique head of
   adductor hallucis m. (cut)
3. Abductor digiti
   minimi m.
4. Lateral plantar nerve
5. Quadratus plantae m. (cut)
6. Flexor digitorum
   brevis m. (cut)
7. Transverse head,
   adductor hallucis m.
8. Common plantar
   digital nerve
9. Medial plantar nerve
10. Calcaneal tuberosity

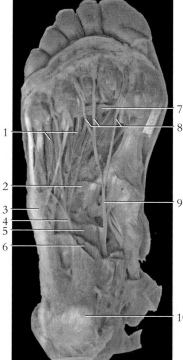

Table 8.13 Muscles of the leg and foot.

Leg Muscle	Origin(s)	Insertion(s)	Action	Innervation
Tibialis anterior	Lateral condyle and body of tibia	First metatarsal bone and first cuneiform bone	Dorsiflexes ankle; inverts foot and ankle	Deep fibular n.
Extensor digitorum longus	Lateral condyle of tibia and anterior surface of fibula	Extensor expansions of digits II–V	Extends digits II–V; dorsiflexes foot at ankle	Deep fibular n.
Extensor hallucis longus	Anterior surface of fibula and interosseous ligament	Distal phalanx of digit I	Extends joints of big toe; assists dorsiflexion of foot at ankle	Deep fibular n.
Fibularis tertius	Anterior surface of fibula and interosseous ligament	Dorsal surface of fifth metatarsal bone	Dorsiflexes and everts foot at ankle	Deep fibular n.
Fibularis longus	Lateral condyle of tibia and head and shaft of fibula	First cuneiform and metatarsal bone I	Plantar flexes and everts foot at ankle	Superficial fibular n.
Fibularis brevis	Lower aspect of fibula	Metatarsal bone V	Plantar flexes and everts foot at ankle	Superficial fibular n.
Gastrocnemius	Lateral and medial condyle of femur	Posterior surface of calcaneous via calcaneal tendon	Plantar flexes foot at ankle; flexes knee joint	Tibial n.
Soleus	Posterior aspect of fibula and tibia	Calcaneous via calcaneal tendon	Plantar flexes foot at ankle	Tibial n.
Plantaris	Supracondylar ridge of femur	Calcaneous	Plantar flexes foot at ankle	Tibial n.
Popliteus	Lateral condyle of femur	Upper posterior aspect of tibia	Flexes and medially rotates leg at knee joint	Tibial n.
Flexor hallucis longus	Posterior surface of tibia	Distal phalanx of big toe	Flexes joint of great toe and plantar flexes foot	Tibial n.
Flexor digitorum longus	Posterior surface of tibia	Distal phalanges of digits II–V	Flexes joints of digits II–V and plantar flexes foot	Tibial n.
Tibialis posterior	Tibia and fibula and interosseous ligament bones II–IV	Navicular, cuneiform, cuboid, and metatarsal bones	Plantar flexes and inverts foot at ankle; supports arches of foot	Tibial n.

Nervous System

The nervous system is anatomically divided into the **central nervous system (CNS)**, which includes the **brain** and **spinal cord**, and the **peripheral nervous system (PNS)**, which includes the **cranial nerves**, arising from the brain, and the **spinal nerves** and ganglia, arising from the spinal cord (Fig. 9.1). The **autonomic nervous system (ANS)** is a functionally distinct division of the nervous system devoted to regulation of involuntary activities in the body. The ANS is comprised of sympathetic and parasympathetic divisions.

The brain and spinal cord are the centers for integration and coordination of information. Conveyed as **nerve impulses**, information to and from the brain and spinal cord travels through **nerves**. Nerves are similar to electrical conducting wires. Nerve impulses are sent from the brain in the form of electrical signals along **motor nerves** to the receiving organs, which then translate the signal into some specific function. For example, the motor impulses conducted from the brain to the muscles of the forearm that serve the hand cause the fingers to move as the muscles are contracted. **Sensory nerves** conduct action potentials (nerve impulses) in the opposite direction–from the receptor site to the CNS. For example, a pinprick on the skin produces a sensory impulse along a sensory nerve that the brain interprets as a painful sensation.

Neurons and **neuroglia** are the two cell types that make up nervous tissue. Neurons are specialized to respond to physical and chemical stimuli, conduct impulses, and release specific chemical regulators, called **neurotransmitters**. Although neurons vary considerably in size and shape, they have three principal components: a **cell body**, **dendrites**, and an **axon** (Fig. 9.3). In a typical neuron connection, the axon of one neuron **synapses** (joins) on the cell body or dendrites of a neighboring neuron. Axons vary in length from a few millimeters in the CNS to over a meter in the PNS. Long axons are generally **myelinated** with **Schwann cells (neurolemmocytes)** in the PNS, and many of the short axons are myelinated with **oligodendrocytes** in the CNS. **Neurofibril nodes (nodes of Ranvier)** are gaps in the **myelin sheath**. The end of the axon at the synapse is called the **axon terminal**.

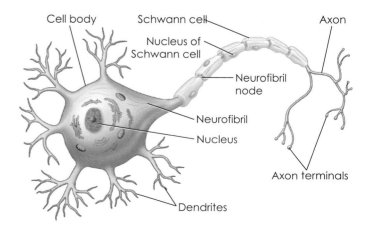

Figure 9.2 A photomicrograph of a neuron.
1. Cytoplasmic extensions   3. Cell body of neuron
2. Nucleus

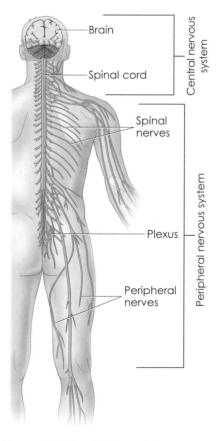

Figure 9.1 The divisions of the nervous system.

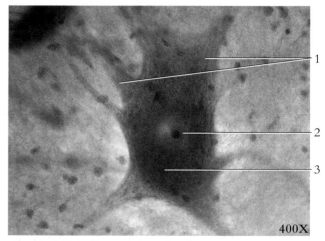

Figure 9.3 The structure of a myelinated neuron.

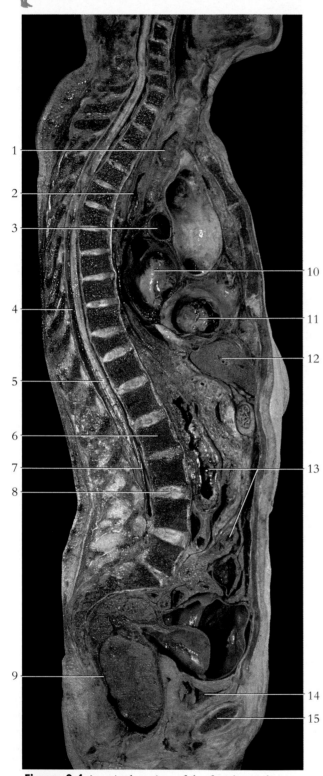

Figure 9.4 A sagittal section of the female trunk.

1. Trachea
2. Esophagus
3. Pulmonary artery
4. Dura mater
5. Spinal cord
6. Body of lumbar vertebra
7. Cauda equina
8. Intervertebral disc
9. Rectum
10. Right atrium
11. Right ventricle
12. Liver
13. Small intestine
14. Urinary bladder
15. Pubic bone

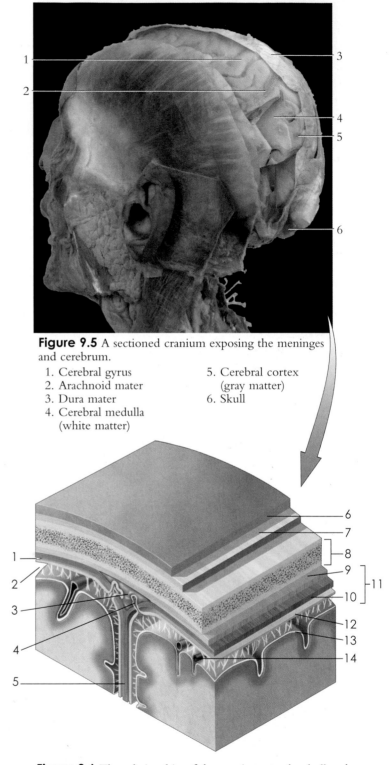

Figure 9.5 A sectioned cranium exposing the meninges and cerebrum.

1. Cerebral gyrus
2. Arachnoid mater
3. Dura mater
4. Cerebral medulla (white matter)
5. Cerebral cortex (gray matter)
6. Skull

Figure 9.6 The relationship of the meninges to the skull and the cerebrum.

1. Subdural space
2. Subarachnoid space
3. Super sagittal sinus
4. Arachnoid granulation
5. Falx cerebri
6. Skin of scalp
7. Periosteum
8. Bone of cranium
9. Periosteal layer
10. Meningeal layer
11. Dura mater
12. Arachnoid mater
13. Pia mater
14. Blood vessel

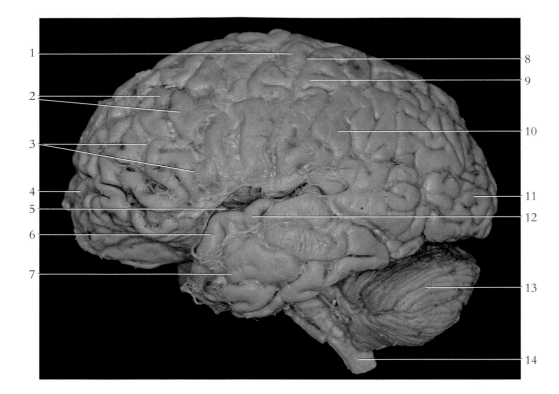

Figure 9.7 A lateral view of the brain.
1. Primary motor cerebral cortex
2. Gyri
3. Sulci
4. Frontal lobe of cerebrum
5. Lateral sulcus
6. Olfactory cerebral cortex
7. Temporal lobe of cerebrum
8. Central sulcus
9. Primary sensory cerebral cortex
10. Parietal lobe of cerebrum
11. Occipital lobe of cerebrum
12. Auditory cerebral cortex
13. Cerebellum
14. Medulla oblongata

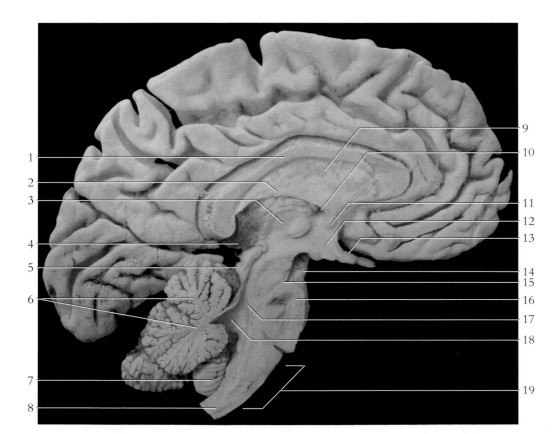

Figure 9.8 A sagittal view of the brain.
1. Body of corpus callosum
2. Crus of fornix
3. Third ventricle covering the thalamus
4. Posterior commissure
5. Inferior colliculus
6. Arbor vitae of cerebellum
7. Tonsilla of cerebellum
8. Spinal cord
9. Septum pellucidum
10. Intraventricular foramen
11. Anterior commissure
12. Hypothalamus
13. Optic chiasma
14. Cerebral peduncle
15. Midbrain
16. Pons
17. Mesencephalic (cerebral) aqueduct
18. Fourth ventricle
19. Medulla oblongata

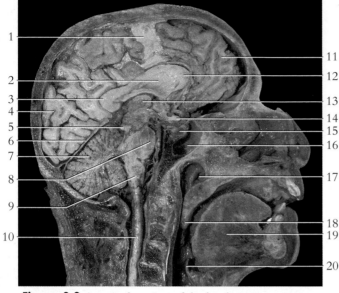

Figure 9.9 A sagittal section of the head.

1. Remnant of falx cerebri
2. Septum pellucidum
3. Genu of corpus callosum
4. Occipital lobe of cerebrum
5. Corpora quadrigemina
6. Tentorium cerebelli
7. Cerebellum
8. Pons
9. Medulla oblongata
10. Spinal cord
11. Frontal lobe of cerebrum
12. Splenium of corpus callosum
13. Thalamus
14. Optic chiasma
15. Pituitary gland
16. Sphenoidal sinus
17. Pharyngeal opening of auditory tube
18. Uvula
19. Tongue
20. Epiglottis

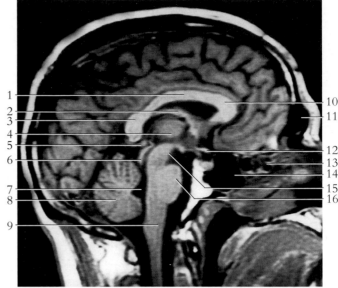

Figure 9.10 An MRI sagittal section through the skull.

1. Body of corpus callosum
2. Fornix
3. Splenium of corpus callosum
4. Thalamus
5. Pineal gland
6. Superior and inferior colliculi
7. Fourth ventricle
8. Cerebellum
9. Medulla oblongata
10. Genu of corpus callosum
11. Frontal sinus
12. Pituitary gland
13. Ethmoidal sinus
14. Sphenoidal sinus
15. Tegmentum (midbrain)
16. Pons

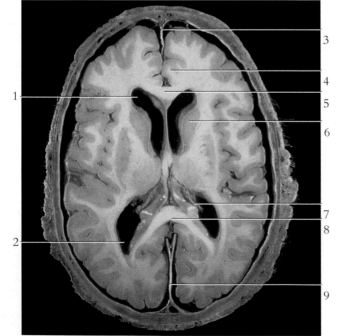

Figure 9.11 A transaxial section of the skull and brain.

1. Anterior horn of lateral ventricle
2. Posterior horn of lateral ventricle
3. Falx cerebri (septum of dura mater)
4. Cingulate gyrus
5. Genu of corpus callosum
6. Caudate nucleus
7. Choroid plexus
8. Splenium of corpus callosum
9. Falx cerebri (septum of dura mater)

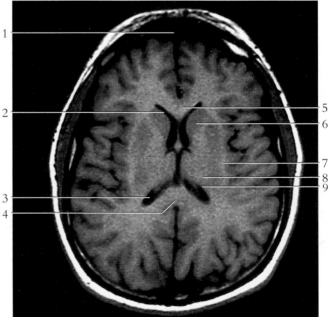

Figure 9.12 An MRI transaxial section through the brain.

1. Frontal sinus
2. Frontal horn of lateral ventricle
3. Posterior horn of lateral ventricle
4. Splenium of corpus callosum
5. Genu of corpus callosum
6. Head of caudate nucleus
7. External capsule
8. Thalamus
9. Choroid plexus

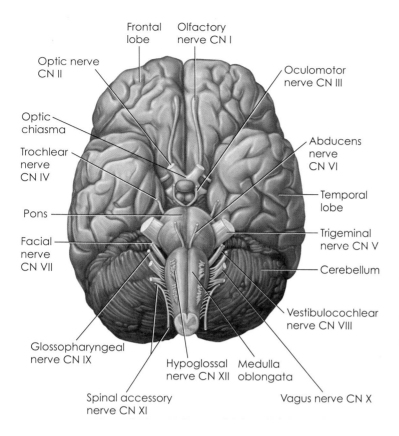

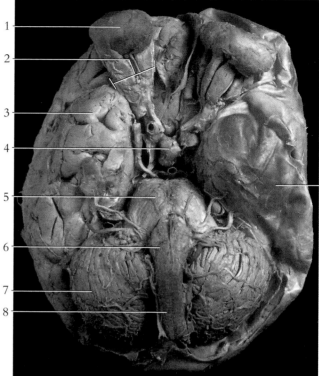

Figure 9.14 An inferior view of the brain with the eyes and part of the meninges still intact.

1. Eyeball
2. Muscles of the eye
3. Temporal lobe of cerebrum
4. Pituitary gland
5. Pons
6. Medulla oblongata
7. Cerebellum
8. Spinal cord
9. Dura mater

Figure 9.13 An illustration of the inferior of the brain showing the cranial nerves.

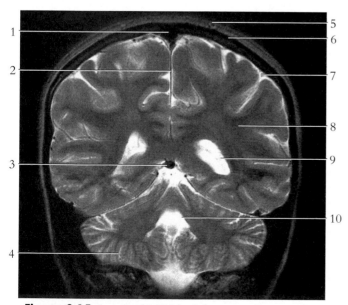

Figure 9.15 A coronal MRI brain scan.
1. Cerebrospinal fluid
2. Longitudinal cerebral fissure
3. Third ventricle
4. Cerebellum
5. Skull
6. Dura mater
7. Cerebral cortex
8. Cerebral medulla
9. Lateral ventricle
10. Fourth ventricle

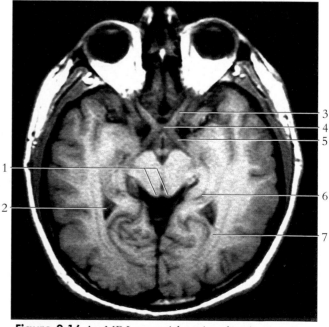

Figure 9.16 An MRI transaxial section showing visual pathways.
1. Superior colliculi
2. Lateral ventricle
3. Optic nerve
4. Optic chiasma
5. Optic tract
6. Lateral geniculate body
7. Calcarine tracts (optic radiation)

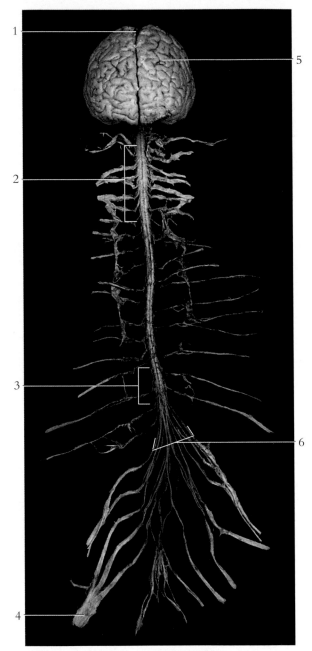

Figure 9.17 The anterior surface of the brain and spinal cord with meninges removed.
1. Longitudinal cerebral fissure
2. Cervical enlargement
3. Lumbar enlargement
4. Sciatic nerve
5. Brain
6. Cauda equina

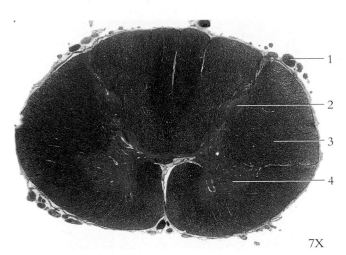

7X

Figure 9.18 Cross section of the spinal cord.
1. Posterior (dorsal) root of spinal nerve
2. Posterior (dorsal) horn (gray matter)
3. Spinal cord tract (white matter)
4. Anterior (ventral) horn (gray matter)

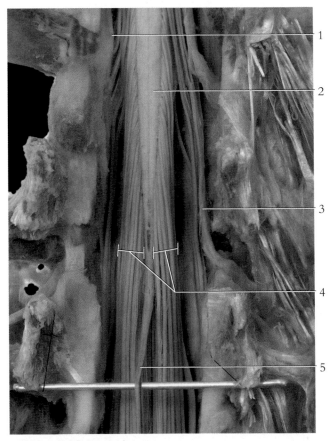

Figure 9.19 A posterior view of the lower spinal cord.
1. Dura mater (cut)
2. Spinal cord
3. Posterior (dorsal) root of spinal nerve
4. Cauda equina
5. Filum terminale

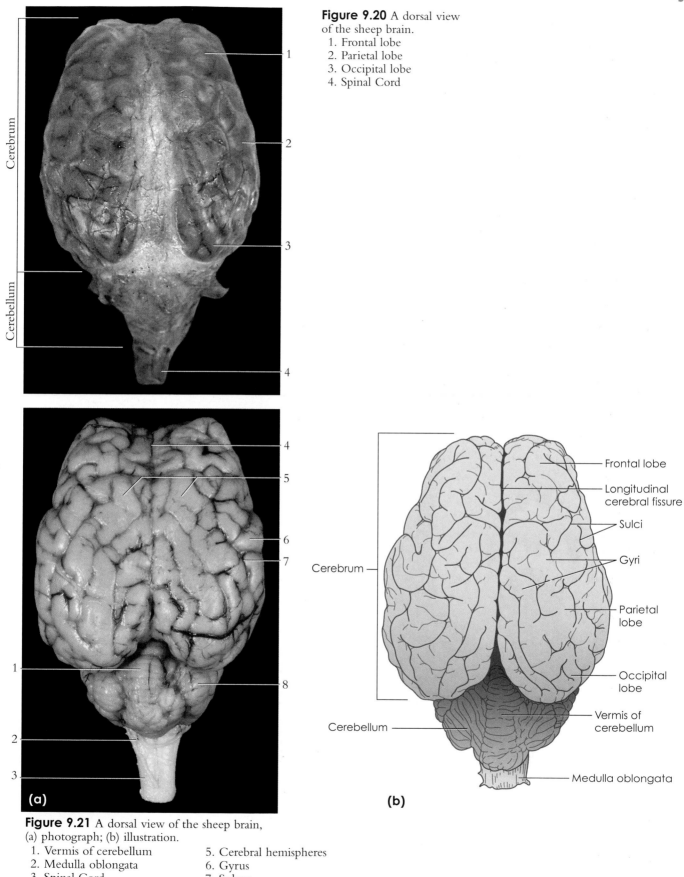

Figure 9.20 A dorsal view of the sheep brain.
1. Frontal lobe
2. Parietal lobe
3. Occipital lobe
4. Spinal Cord

Figure 9.21 A dorsal view of the sheep brain,
(a) photograph; (b) illustration.
1. Vermis of cerebellum
2. Medulla oblongata
3. Spinal Cord
4. Longitudinal cerebral fissure
5. Cerebral hemispheres
6. Gyrus
7. Sulcus
8. Cerebellar hemisphere

111

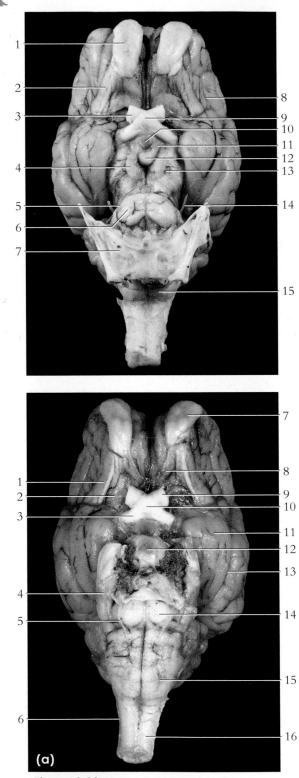

Figure 9.22 A ventral view of sheep brain with dura mater cut and reflected.

1. Olfactory bulb
2. Olfactory tract
3. Optic nerve
4. Oculomotor nerve
5. Trigeminal nerve
6. Pons
7. Dura mater (cut)
8. Pia mater (adhering to brain)
9. Optic chiasma
10. Position of pituitary stock
11. Tuber cinereum
12. Mammillary body
13. Cerebral penduncle
14. Trochlear nerve
15. Medulla oblongata

Figure 9.23 A ventral view of sheep brain, (a) photograph and (b) illustration.

1. Lateral olfactory band
2. Olfactory trigone
3. Optic tract
4. Trigeminal nerve
5. Abducens nerve
6. Accessory nerve
7. Olfactory bulb
8. Medial olfactory band
9. Optic nerve
10. Optic chiasma
11. Pyriform lobe
12. Pituitary gland (hypophysis)
13. Rhinal sulcus
14. Pons
15. Medulla oblongata
16. Spinal cord

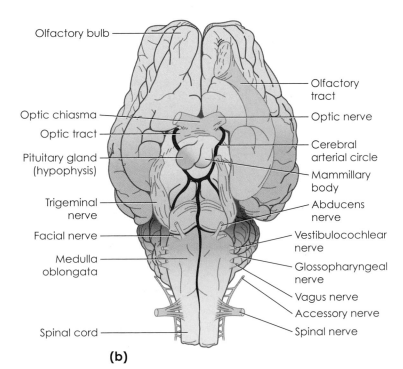

(b)

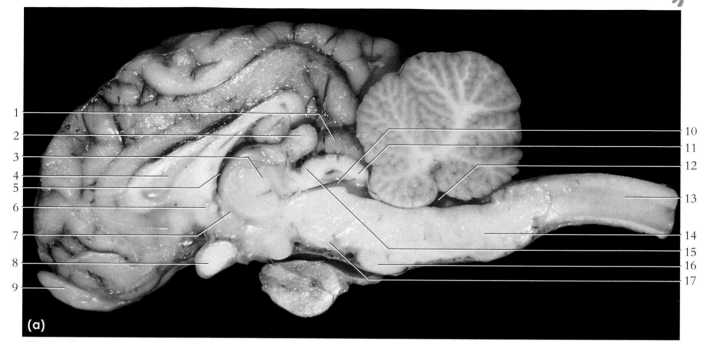

(a)

Figure 9.24 A right sagittal view of the sheep brain, (a) photograph and (b) illustration.

1. Superior colliculus
2. Pineal body (gland)
3. Intermediate mass
4. Septum pellucidum
5. Interventicular foramen (foramen of Monro)
6. Anterior commissure
7. Third ventricle
8. Optic chiasma
9. Olfactory bulb
10. Mesencephalic (cerebral) aqueduct
11. Inferior colliculus
12. Fourth ventricle
13. Spinal cord
14. Medulla oblongata
15. Posterior commissure
16. Pons
17. Cerebral peduncle

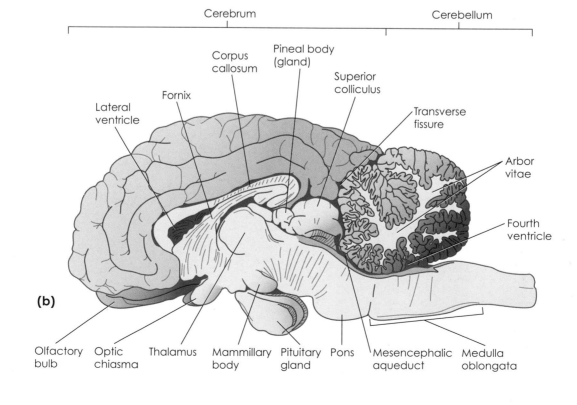

(b)

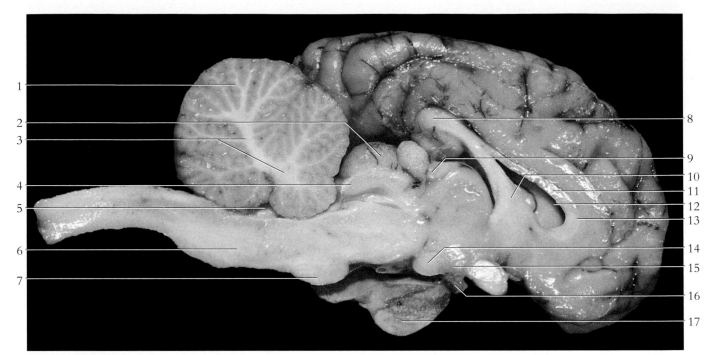

Figure 9.25 A left sagittal view of the sheep brain.

1. Cerebellum
2. Superior colliculus
3. Arbor vitae
4. Inferior colliculus
5. Fourth ventricle
6. Medulla oblongata

7. Pons
8. Splenium of corpus callosum
9. Habenular trigone
10. Fornix
11. Body of corpus callosum
12. Lateral ventricle

13. Genu of corpus callosum
14. Mammillary body
15. Tuber cinereum
16. Pituitary stalk
17. Pituitary gland (hypophysis)

Figure 9.26 A lateral view of the brainstem.
1. Pons
2. Abducens nerve
3. Medulla oblongata
4. Hypoglossal nerve
5. Spinal cord
6. Lateral geniculate body
7. Medial geniculate body
8. Trochlear nerve
9. Trigeminal nerve
10. Accessory nerve

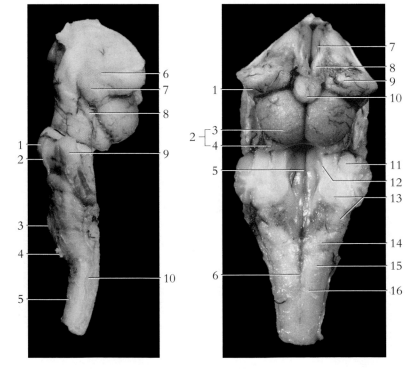

Figure 9.27 A dorsal view of the brainstem.
1. Medial geniculate body
2. Corpora quadrigemina
3. Superior colliculus
4. Inferior colliculus
5. Fourth ventricle
6. Dorsal median sulcus
7. Intermediate mass
8. Habenular trigone
9. Thalmus
10. Pineal gland
11. Middle cerebellar peduncle
12. Anterior cerebellar peduncle
13. Posterior cerebellar peduncle
14. Tuberculum cuneatum
15. Fasciculus gracilis
16. Fasciculus cuneatus

Endocrine System

Chapter 10

The endocrine system works closely with the nervous system to regulate and integrate body processes and maintain homeostasis. The nervous system regulates body activities through the action of electrochemical impulses that are transmitted by means of neurons, resulting in rapid, but usually brief responses. By contrast, the endocrine system is composed of glands (Fig. 10.1) scattered throughout the body that release chemical substances called **hormones** into the bloodstream. These hormones dissipate in the blood and travel throughout the entire body to act on **target tissues**, where they have a slow but relatively long-lasting effect. Neurological responses are measured in milliseconds, but hormonal action requires seconds or days to elicit a response. Some hormones may have an effect that lasts for minutes and others for weeks or months.

The endocrine system and nervous system are closely coordinated in autonomically controlling the functions of the body. The **pituitary gland**, located next to the brain, regulates the activity of most other endocrine glands. Located immediately between the pituitary and the rest of the brain is the **hypothalamus**. The hypothalamus serves as an intermediate between the nervous centers of the brain and the pituitary gland, coordinating the activity of the two systems. Furthermore, certain hormones may stimulate or inhibit the activities of the nervous system.

Other organs of the endocrine system include the **thyroid gland** and **parathyroid glands**, located in the neck. The **adrenal glands** and **pancreas** are located in the abdominal region. The **ovaries** of the female are located in the pelvic cavity, whereas the **testes** of the male are located in the scrotum. Both of their reproductive and endocrine roles will be discussed in Chapter 17. Even the **placenta** serves as an endocrine organ for the developing fetus and has some hormonal influence upon the pregnant woman.

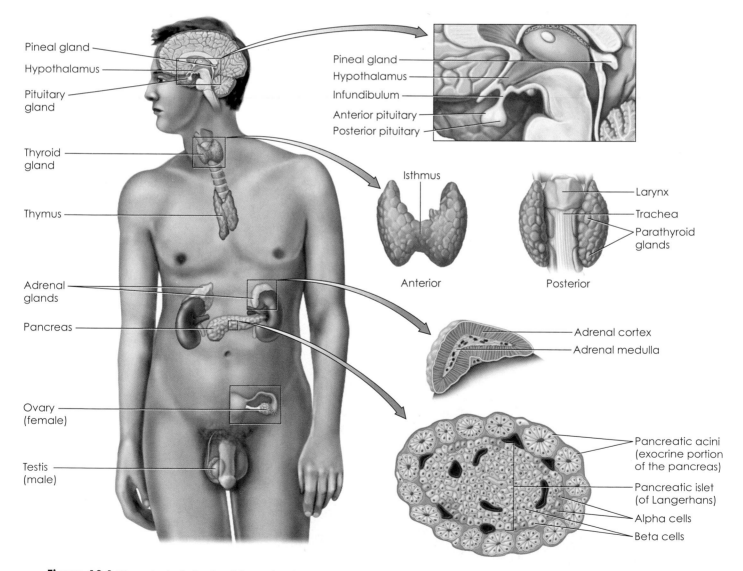

Figure 10.1 The principal glands of the endocrine system.

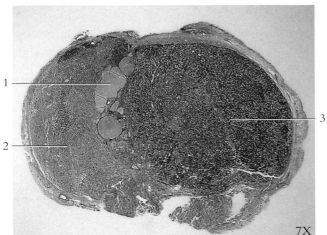

Figure 10.2 Pituitary gland.
1. Pars intermedia (adenohypophysis)
2. Pars nervosa (neurohypophysis)
3. Pars distalis (adenohypophysis)

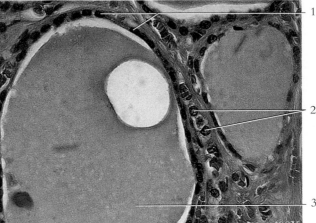

Figure 10.3 Thyroid gland.
1. Follicle cells
2. C cells
3. Colloid within follicle

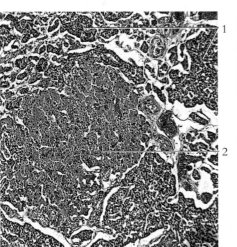

Figure 10.4 Parathyroid gland.
1. Chief cells
2. Cluster of oxyphil cells

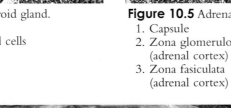

Figure 10.5 Adrenal gland.
1. Capsule
2. Zona glomerulosa (adrenal cortex)
3. Zona fasiculata (adrenal cortex)
4. Zona reticularis (adrenal cortex)
5. Adrenal medulla

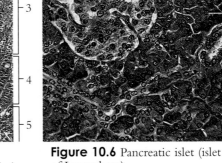

Figure 10.6 Pancreatic islet (islet of Langerhans).
1. Pancreatic islet (endocrine pancreas)
2. Acini (exocrine pancreas)
3. Pancreatic duct (exocrine pancreas)

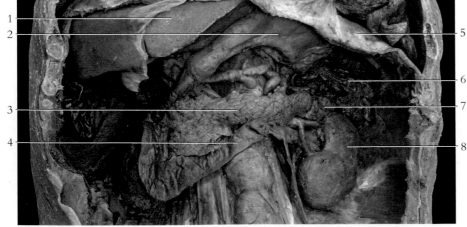

Figure 10.7 The adrenal (suprarenal) gland and pancreas with associated structures within the abdominal cavity with overlying viscera removed.
1. Liver (reflected)
2. Stomach (reflected)
3. Pancreas
4. Duodenum
5. Diaphragm
6. Spleen
7. Adrenal gland
8. Kidney

116

Sensory Organs

The nervous and endocrine systems convey information from the brain to all parts of the body to enable a person to interact with both the external and internal environments and to maintain homeostasis. The sense organs, in contrast, convey information from the outside world (and inside world of the body) back to the brain. This includes a wide range of information such as temperature, brightness, sound, flavor, smell, and balance.

The sense organs are actually extensions of the nervous system that allow us to autonomically respond to, or conscientiously perceive, our internal and external environments. A stimulus excites a sense organ which then transduces the stimulus to an electrical (nerve) impulse. Sensory nerves transmit the impulse (sensation) to the brain to be perceived and acted upon. Ultimately, it is the brain which actually feels, sees, hears, tastes, and smells.

Taste sensation is produced by chemicals from food binding to taste receptors on thousands of taste buds located on fungiform and circumvallate papillae.

The **ear** is the organ of hearing and equilibrium (balance). It contains receptors that respond to movements of the head and receptors that convert sound waves into nerve impulses. Impulses from both receptor types are transmitted through the vestibulocochlear (VIII) cranial nerve to the brain for interpretation.

The ear consists of three principal regions: the **outer ear**, the **middle ear**, and the **inner ear**. The outer ear consists of the **auricle** and the **external auditory canal**. The middle ear contains tympanic membrane and the auditory ossicles (**malleus, incus, and stapes**). The inner ear contains the **spiral organ** (organ of Corti) in the **cochlea** for hearing, and the **semicircular canals** and the vestibular organs for equilibrium.

The **eyes** are the organs of visual sense. The eyes refract (bend) and focus the incoming light waves onto the sensitive **photoreceptors** (**rods and cones**) at the back of each eye. Nerve impulses from the stimulated photoreceptors are conveyed along visual pathways to the occipital lobes of the cerebrum, where visual sensations are perceived.

The eyeball consists of the fibrous tunic, which is divided into the **sclera** and **cornea**; the vascular tunic, which consists of the **choroid**, the **ciliary body**, and the **iris**; and the internal tunic, or **retina**, which consists of an outer pigmented layer and an inner nervous layer. The eye contains an anterior cavity between the lens and the cornea. The anterior cavity is subdivided into an anterior chamber in front of the iris and a posterior chamber behind the iris. **Aqueous humor** fills both of these chambers. The posterior cavity (also called the vitreous chamber) contains vitreous humor and is located between the lens and the retina.

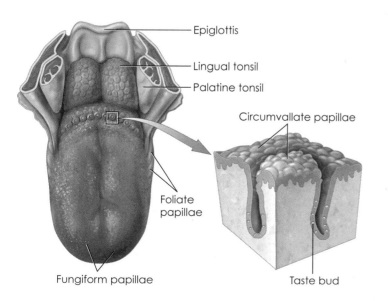

Figure 11.1 The surface of the tongue.

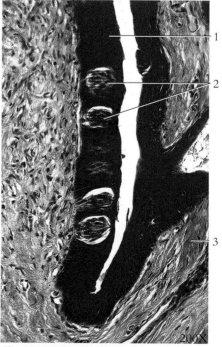

Figure 11.2 A taste bud.
1. Epithelium of papilla
2. Taste buds
3. Tongue muscle

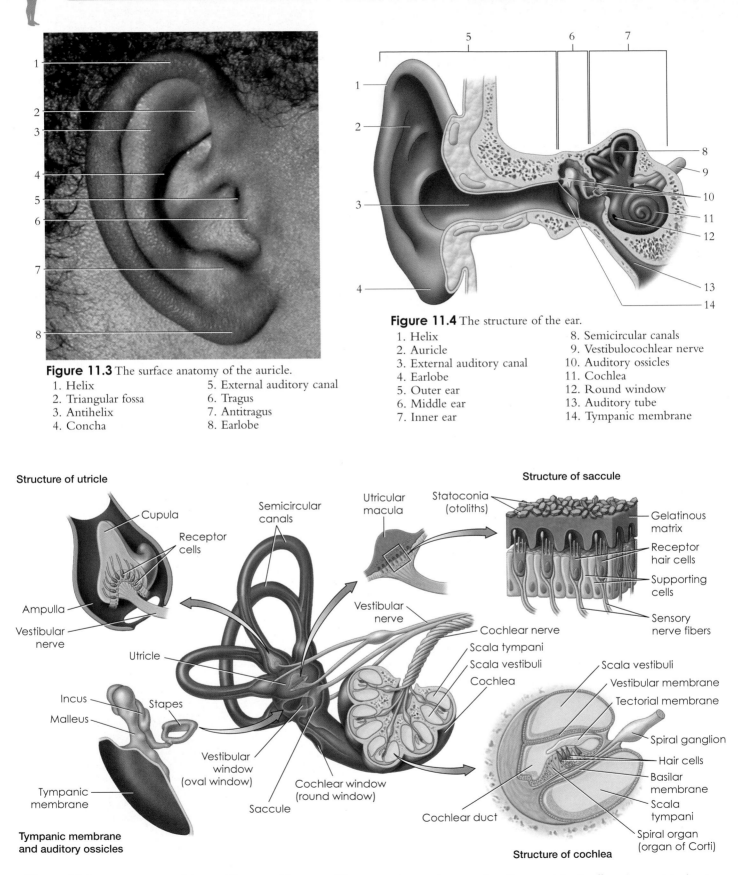

Figure 11.3 The surface anatomy of the auricle.

1. Helix
2. Triangular fossa
3. Antihelix
4. Concha
5. External auditory canal
6. Tragus
7. Antitragus
8. Earlobe

Figure 11.4 The structure of the ear.

1. Helix
2. Auricle
3. External auditory canal
4. Earlobe
5. Outer ear
6. Middle ear
7. Inner ear
8. Semicircular canals
9. Vestibulocochlear nerve
10. Auditory ossicles
11. Cochlea
12. Round window
13. Auditory tube
14. Tympanic membrane

Figure 11.5 The structures of the middle ear and inner ear. The tympanic membrane and auditory ossicles (malleus, incus, stapes) are structures of the middle ear. The vestibular organs (utricle, saccule, semicircular canals) and cochlea (containing the spiral organ) are structures of the inner ear.

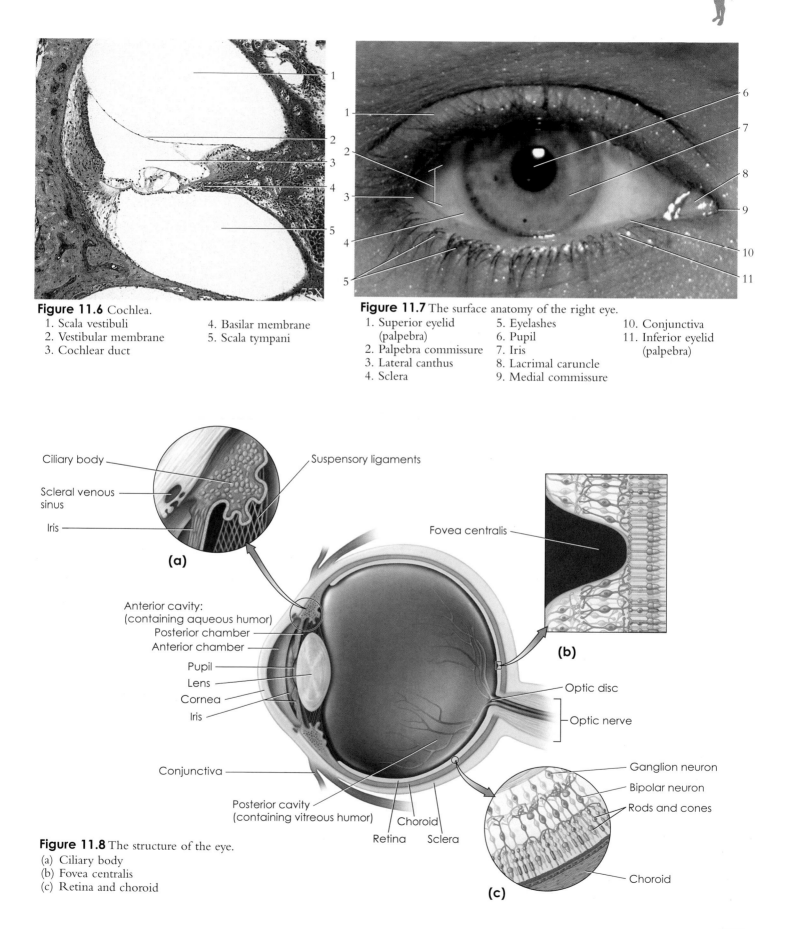

Figure 11.6 Cochlea.
1. Scala vestibuli
2. Vestibular membrane
3. Cochlear duct
4. Basilar membrane
5. Scala tympani

Figure 11.7 The surface anatomy of the right eye.
1. Superior eyelid (palpebra)
2. Palpebra commissure
3. Lateral canthus
4. Sclera
5. Eyelashes
6. Pupil
7. Iris
8. Lacrimal caruncle
9. Medial commissure
10. Conjunctiva
11. Inferior eyelid (palpebra)

Figure 11.8 The structure of the eye.
(a) Ciliary body
(b) Fovea centralis
(c) Retina and choroid

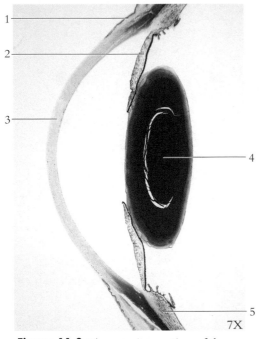

Figure 11.9 The anterior portion of the eye.
1. Conjunctiva   4. Lens
2. Iris   5. Ciliary body
3. Cornea

7X

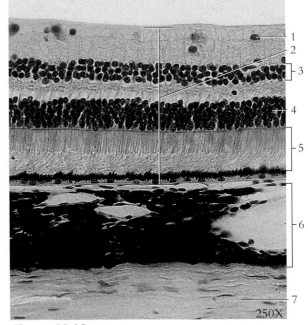

Figure 11.10 The retina.
1. Ganglion nucleus   5. Rods and cones
2. Retina   6. Choroid
3. Bipolar nuclei   7. Sclera
4. Rods and cones nuclei

250X

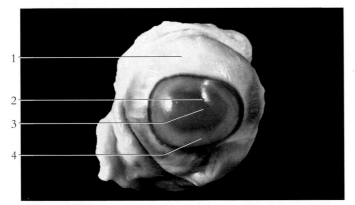

Figure 11.11 A superficial view of the anterior eyeball of a sheep.
1. Sclera   3. Pupil (dark opening)
2. Cornea   4. Iris

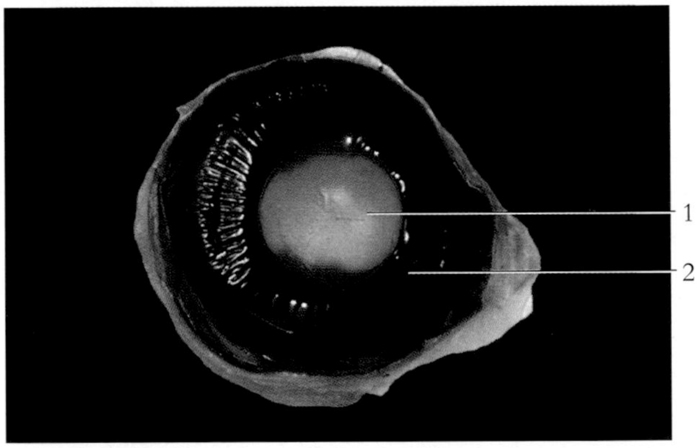

Figure 11.12 An anterior view of the eyeball with the lens in natural position.
1. Lens   2. Iris

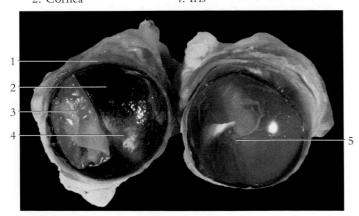

Figure 11.13 Tunics of the sheep eyeball.
1. Sclera   4. Tapetum lucidum
2. Choroid   5. Optic disc
3. Retina

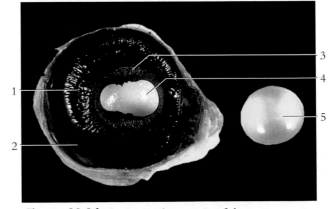

Figure 11.14 The internal anatomy of sheep eye.
1. Ciliary body   4. Pupil
2. Ora serrata   5. Lens (removed)
3. Iris

Circulatory System

Chapter 12

The circulatory system consists of the **blood**, **heart**, and **vessels**, each of which is essential to the life of a complex multicellular organism. Blood, a specialized connective tissue, consists of **formed elements** (erythrocytes, leukocytes, and thrombocytes) that are suspended and carried in the **plasma**. These formed elements function in transport, immunity, and blood-clotting mechanisms.

The heart is enclosed in a **pericardial sac** within the thoracic cavity. The wall of the heart consists of the **epicardium**, **myocardium**, and **endocardium**. The **right atrium** of the heart receives deoxygenated blood from the superior vena cava and inferior vena cava, and the **right ventricle** pumps deoxygenated blood into the **pulmonary trunk** to the **pulmonary arteries**. The **left atrium** receives oxygenated blood from the **pulmonary veins** and pumps oxygenated blood into the **left ventricle**. The left ventricle pumps blood into the **aorta**.

There are four heart valves that prohibit the backflow of blood: 1) The **right atrioventricular valve (tricuspid valve)** is located between the right atrium and the right ventricle; 2) the **pulmonary valve (pulmonary semilunar valve)** is located between the right ventricle and the pulmonary trunk; 3) the **left atrioventricular valve (bicuspid, or mitral valve)** is located between the left atrium and the left ventricle; and, 4) the **aortic valve (aortic semilunar valve)** is located between the left ventricle and the ascending aorta.

The **systemic arteries** arise from the aorta or branches of the aorta and transport blood away from the heart to smaller vessels called **arterioles**. From arterioles, the blood enters **capillaries** where diffusion with the surrounding cells may occur. Capillaries converge forming **venules**, which in turn converge forming larger vessels called **veins**. Veins are vessels that transport blood toward the heart.

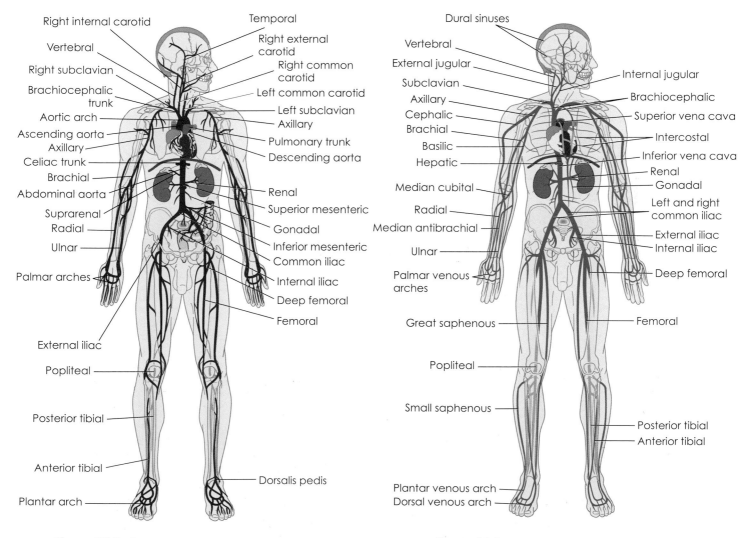

Figure 12.1 The principal arteries of the body.

Figure 12.2 The principal veins of the body.

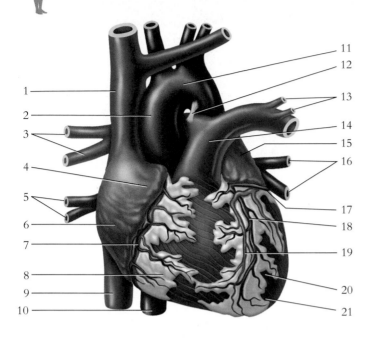

Figure 12.3 An anterior view of the structure of the heart.

1. Superior vena cava
2. Ascending aorta
3. Branches of right pulmonary artery
4. Auricle of right atrium
5. Right pulmonary veins
6. Right atrium
7. Right coronary artery and vein
8. Right ventricle
9. Inferior vena cava
10. Thoracic aorta
11. Aortic arch
12. Ligamentum arteriosum
13. Branches of left pulmonary artery
14. Pulmonary trunk
15. left atrium
16. Left pulmonary veins
17. Circumflex artery
18. Anterior interventricular artery
19. Anterior interventricular vein
20. Left ventricle
21. Apex of heart

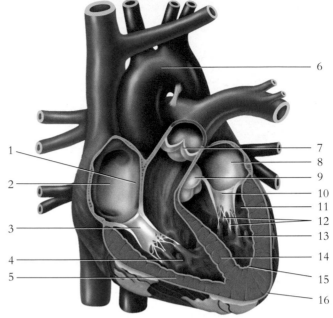

Figure 12.4 An internal view of the structure of the heart.

1. Interatrial septum
2. Right atrium
3. Tricuspid valve
4. Right ventricle
5. Myocardium
6. Aortic arch
7. Pulmonary valve
8. Left atrium
9. Aortic valve
10. Bicuspid valve
11. Left ventricle
12. Chordae tendinae
13. Papillary muscle
14. Interventricular septum
15. Endocardium
16. Visceral pericardium

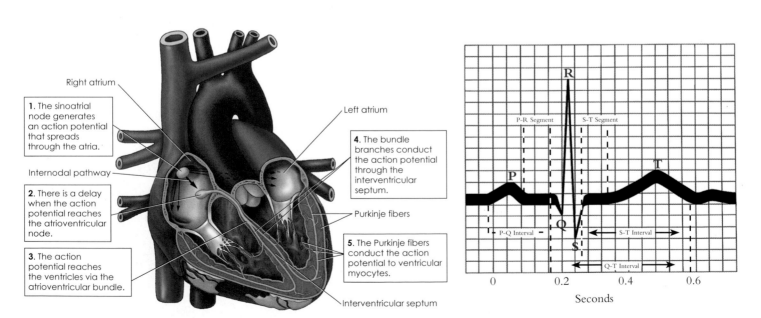

Figure 12.5 The conduction system of the heart.

Right atrium

1. The sinoatrial node generates an action potential that spreads through the atria.

Internodal pathway

2. There is a delay when the action potential reaches the atrioventricular node.

3. The action potential reaches the ventricles via the atrioventricular bundle.

Left atrium

4. The bundle branches conduct the action potential through the interventricular septum.

Purkinje fibers

5. The Purkinje fibers conduct the action potential to ventricular myocytes.

Interventricular septum

Figure 12.6 A normal electrocardiogram (ECG).

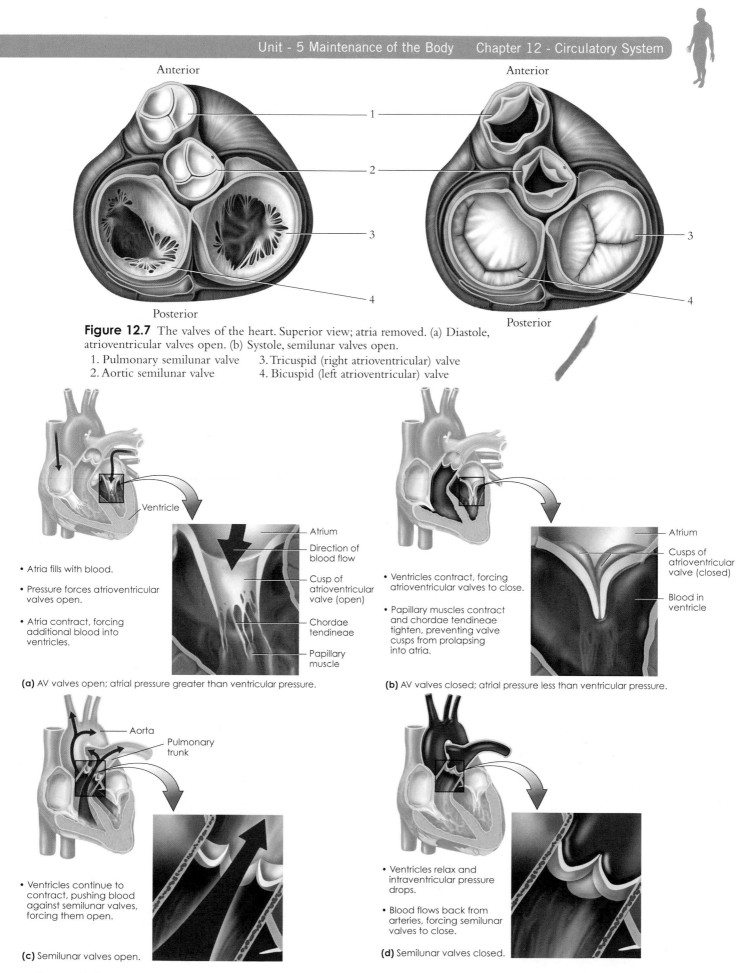

Figure 12.7 The valves of the heart. Superior view; atria removed. (a) Diastole, atrioventricular valves open. (b) Systole, semilunar valves open.

1. Pulmonary semilunar valve   3. Tricuspid (right atrioventricular) valve
2. Aortic semilunar valve   4. Bicuspid (left atrioventricular) valve

• Atria fills with blood.

• Pressure forces atrioventricular valves open.

• Atria contract, forcing additional blood into ventricles.

Atrium

Direction of blood flow

Cusp of atrioventricular valve (open)

Chordae tendineae

Papillary muscle

Ventricle

(a) AV valves open; atrial pressure greater than ventricular pressure.

• Ventricles contract, forcing atrioventricular valves to close.

• Papillary muscles contract and chordae tendineae tighten, preventing valve cusps from prolapsing into atria.

Atrium

Cusps of atrioventricular valve (closed)

Blood in ventricle

(b) AV valves closed; atrial pressure less than ventricular pressure.

Aorta

Pulmonary trunk

• Ventricles continue to contract, pushing blood against semilunar valves, forcing them open.

(c) Semilunar valves open.

• Ventricles relax and intraventricular pressure drops.

• Blood flows back from arteries, forcing semilunar valves to close.

(d) Semilunar valves closed.

Figure 12.8 Cardiac cycle.

123

Table 12.1 Valves of the heart.

Valve	Location	Structure and Function
Right atrioventricular valve (tricuspid valve)	Between right atrium and right ventricle	Consists of three cusps that prevent backflow of blood into right atrium during systole (ventricular contraction)
Pulmonary valve (pulmonary semilunar valve)	Entrance to pulmonary trunk	Consists of three partial moon-shaped cups that prevent backflow of blood into right ventricle during diastole (ventricular relaxation)
Left atrioventricular valve (bicuspid valve or mitral valve)	Between left atrium and left ventricle	Consists of two cusps that prevent backflow of blood into left atrium during systole
Aortic valve (aortic semilunar valve)	Entrance to ascending aorta	Consists of three partial moon-shaped cups that prevent backflow of blood into left ventricle during diastole

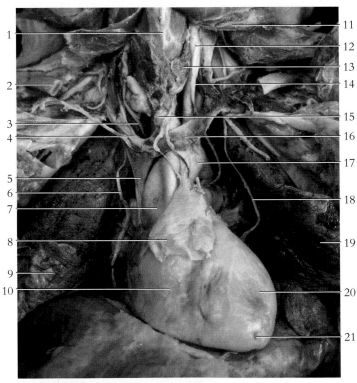

Figure 12.9 An anterior view of the heart and associated structures.

1. Thyroid cartilage of larynx	12. Left common carotid artery
2. First rib (cut)	13. Thyroid gland (cut)
3. Right vagus nerve	14. Left vagus nerve
4. Right brachiocephalic vein	15. Brachiocephalic artery
5. Superior vena cava	16. Left brachiocephalic vein
6. Right phrenic nerve	17. Aortic arch
7. Ascending aorta	18. Left phrenic nerve
8. Pericardium (cut)	19. Left lung
9. Right lung	20. Left ventricle of heart
10. Right ventricle of heart	21. Apex of heart
11. Sternohyoid (cut and reflected)	

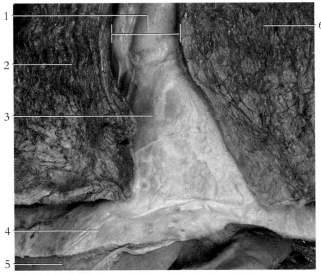

Figure 12.10 The position of the heart within the pericardium.

1. Mediastinum	4. Diaphragm
2. Right lung	5. Liver
3. Pericardium	6. Left lung

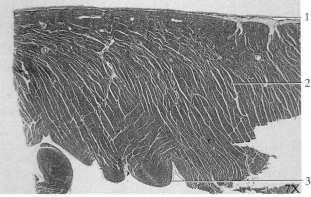

Figure 12.11 Wall of the heart.

1. Epicardium	3. Endocardium
2. Myocardium	

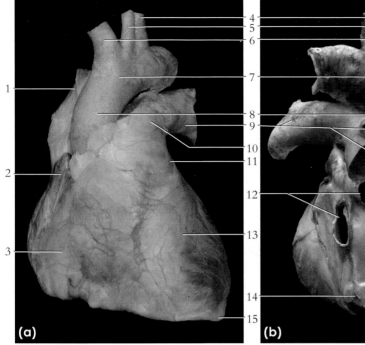

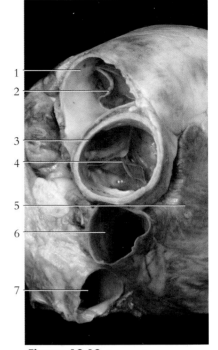

Figure 12.12 An anterior (a) and posterior (b) views of the heart and great vessels.
1. Superior vena cava
2. Right atrium
3. Right ventricle
4. Left subclavian artery
5. Left common carotid artery
6. Brachiocephalic trunk
7. Aortic arch
8. Ascending aorta
9. Pulmonary arteries
10. Pulmonary trunk
11. Left atrium
12. Pulmonary veins
13. Left ventricle
14. Inferior vena cava
15. Apex of heart
16. Superior vena cava
17. Right atrium
18. Right ventricle

Figure 12.13 A superior view of the great vessels of the heart and aortic valve.
1. Pulmonary trunk
2. Pulmonary valve
3. Ascending aorta
4. Aortic valve
5. Right atrium
6. Superior vena cava
7. Pulmonary vein

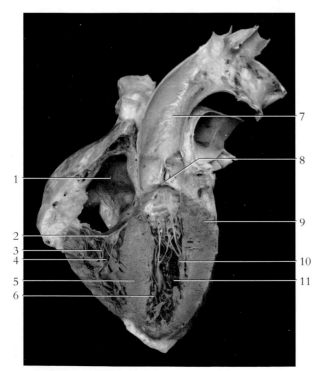

Figure 12.14 The internal structure of the heart.
1. Right atrium
2. Right atrioventricular valve
3. Chordae tendinae
4. Right ventricle
5. Interventricular septum
6. Trabeculae carneae
7. Ascending aorta
8. Aortic valve
9. Myocardium
10. Papillary muscle
11. Left ventricle

Figure 12.15 A double coronary artery bypass surgery. Several vessels may be used in the autotransplant, including the internal thoracic artery and the great saphenous vein.
1. A graft to the ascending aorta
2. A graft to the left coronary artery

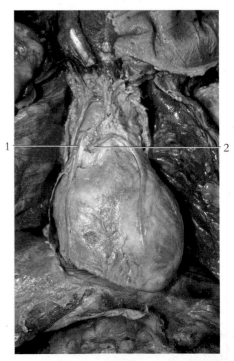

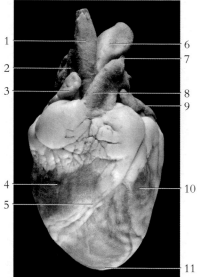

Figure 12.16 A ventral view of mammalian (sheep) heart.

1. Brachiocephalic artery
2. Cranial vena cava
3. Right auricle of right atrium
4. Right ventricle
5. Interventicular groove
6. Aortic arch
7. Ligamentum arteriosum
8. Pulmonary trunk
9. Left auricle of left atrium
10. Left ventricle
11. Apex of heart

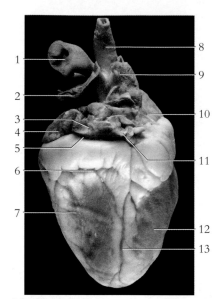

Figure 12.17 A dorsal view of mammalian (sheep) heart.

1. Aorta
2. Pulmonary artery
3. Pulmonary vein
4. Left auricle
5. Left atrium
6. Atrioventricular groove
7. Left ventricle
8. Brachiocephalic artery
9. Cranial vena cava
10. Right auricle
11. Pulmonary vein
12. Right ventricle
13. Interventricular groove

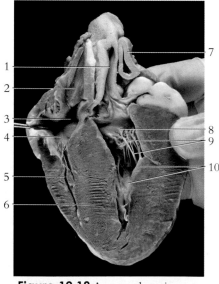

Figure 12.18 A coronal section of the mammalian (sheep) heart .

1. Aorta
2. Cranial vena cava
3. Right atrium
4. Right atrioventricular (tricuspid) valve
5. Right ventricle
6. Interventricular septum
7. Pulmonary artery
8. Left atrioventricular (bicuspid) valve
9. Chordae tendineae
10. Papillary muscles

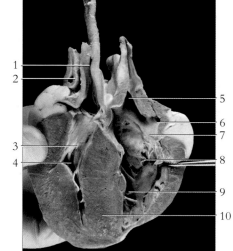

Figure 12.19 A coronal section of the mammalian (sheep) heart showing the valves.

1. Opening of the brachiocephalic artery
2. Pulmonary artery
3. Left atrioventricular (bicuspid) valve
4. Left ventricle
5. Opening of cranial vena cava
6. Opening of coronary sinus
7. Right atrium
8. Right atrioventricular (tricuspid) valve
9. Right ventricle
10. Interventricular septum

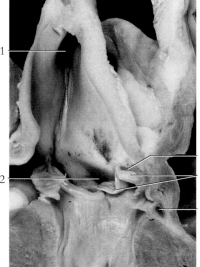

Figure 12.20 A coronal section of the mammalian (sheep) heart showing openings of coronary arteries.

1. Opening of brachiocephalic artery
2. Opening of left coronary artery
3. Opening of right coronary artery
4. Aortic valve
5. Coronary vessel

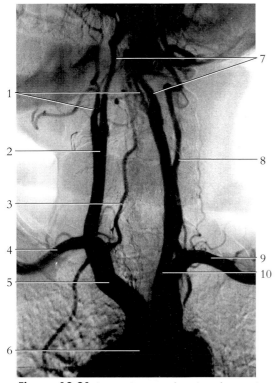

Figure 12.21 An angiogram showing the aortic arch and its branches.

1. External carotid arteries
2. Right common carotid artery
3. Right vertebral artery
4. Right subclavian artery
5. Brachiocephalic trunk
6. Aortic arch
7. Internal carotid arteries
8. Left vertebral artery
9. Left subclavian artery
10. Left common carotid artery

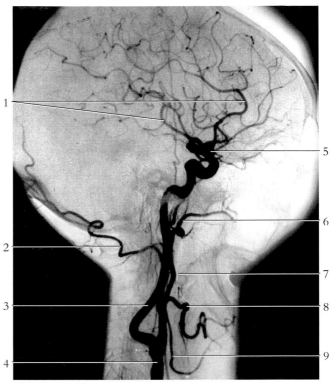

Figure 12.22 An angiogram showing the branches of the common carotid and external carotid arteries.

1. Meningeal arteries
2. Occipital artery
3. Internal carotid artery
4. Common carotid artery
5. Internal carotid artery to cerebral arterial circle (circle of Willis)
6. Maxillary artery
7. External carotid artery
8. Facial artery
9. Superior thyroid artery

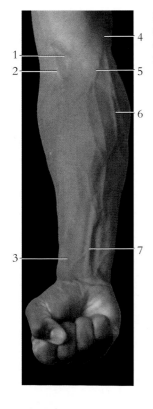

Figure 12.23 The surface anatomy identifying the superficial vessels of the forearm.

1. Cubital fossa
2. Cephalic vein
3. Radial artery (arterial pressure point)
4. Basilic vein
5. Median cubital vein
6. Median antebrachial vein
7. Tendon of palmaris longus m.

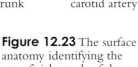

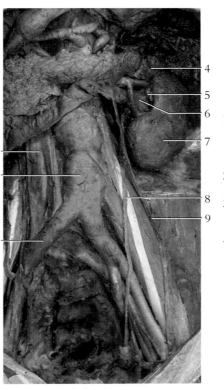

Figure 12.24 The vessels of the lower abdominal cavity.

1. Inferior vena cava
2. Abdominal aorta
3. Right common iliac artery
4. Adrenal gland
5. Left renal artery
6. Left renal vein
7. Left kidney
8. Ureter
9. Left gonadal vein

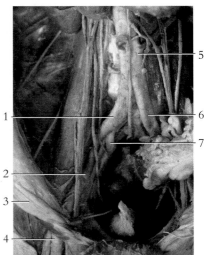

Figure 12.25 The arteries of the pelvic cavity.

1. Right common iliac artery
2. External iliac artery
3. Inguinal ligament
4. Femoral artery
5. Abdominal aorta
6. Left common iliac artery
7. Internal iliac artery

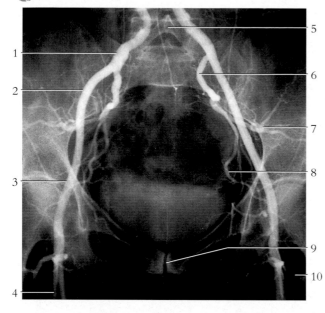

Figure 12.26 An angiogram of the common iliac arteries and their branches.
1. Common iliac artery
2. External iliac artery
3. Femoral artery
4. Deep femoral artery
5. Lumbar vertebra
6. Internal iliac artery
7. Gluteal arteries
8. Obturator artery
9. Symphysis pubis
10. Lateral circumflex femoral artery

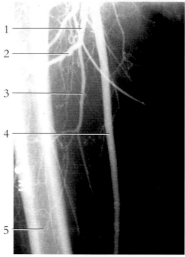

Figure 12.27 An angiogram of the arteries of the right thigh.
1. Deep femoral artery
2. Lateral circumflex femoral artery
3. Medial femoral circumflex artery
4. Femoral artery
5. Femur

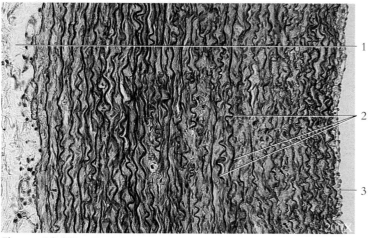

Figure 12.28 Wall of elastic artery.
1. Tunica adventitia
2. Elastic laminae (in tunica media)
3. Tunica intima

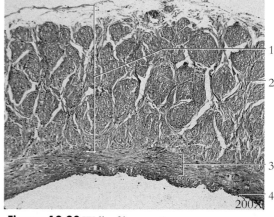

Figure 12.29 Wall of large vein.
1. Tunica adventitia
2. Longitudinally oriented smooth muscle
3. Tunica media
4. Tunica intima

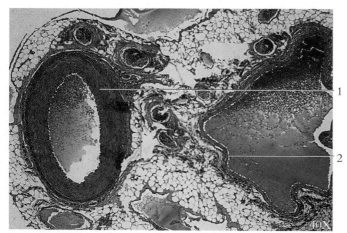

Figure 12.30 Comparison of artery and vein structure.
1. Artery
2. Vein

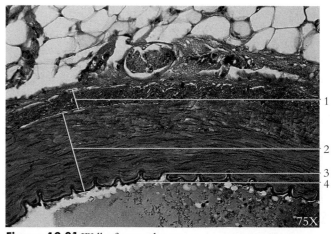

Figure 12.31 Wall of muscular artery.
1. Tunica adventitia
2. Tunica media
3. Internal elastic membrane
4. Endothelial cell

128

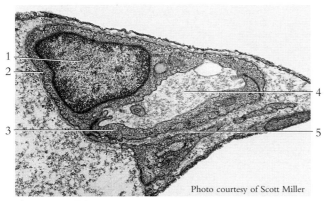

Figure 12.32 SEM photomicrograph of a capillary.
1. Endothelial cell nucleus
2. Endocytic vesicles
3. Endothelial cell
4. Lumen of capillary
5. Basal lamina

Photo courtesy of Scott Miller

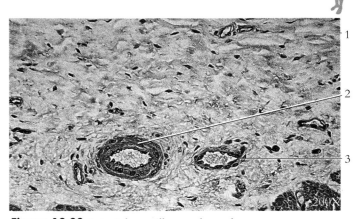

Figure 12.33 Arteriole, capillary, and venule.
1. Capillary
2. Arteriole
3. Venule

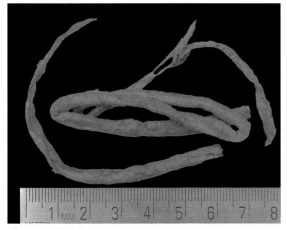

Figure 12.34 Arterial plaque from femoral arteries.

Figure 12.35 An electron micrograph of blood cells in the lumen of a blood vessel.
1. Leukocytes
2. Erythrocytes

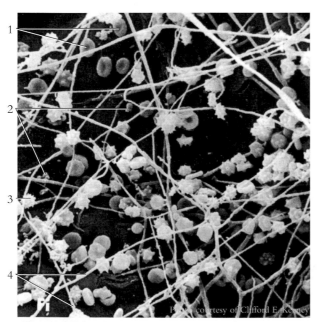

Figure 12.36 An electron micrograph of a blood clot.
1. Erythrocytes
2. Thrombocytes
3. Leukocyte
4. Fibrin strand

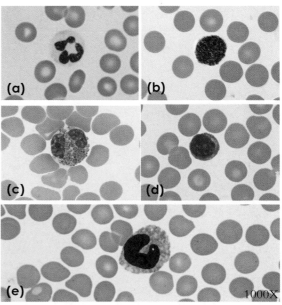

Figure 12.37 Types of leukocytes.
(a) Neutrophil
(b) Basophil
(c) Eosinophil
(d) Lymphocyte
(e) Monocyte

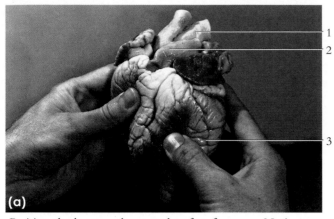

(a)

Position the heart so the ventral surface faces you. Notice the thicker ventricular walls, especially the left ventricle.
1. Aortic arch
2. Pulmonary trunk
3. Left ventricle

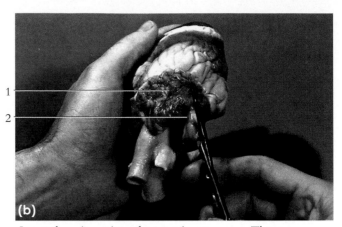

(b)

Insert the scissors into the superior vena cava. The cut should expose the interior of the right atrium. Notice the right atrioventricular (tricuspid) valve.
1. Right atrium
2. Superior vena cava

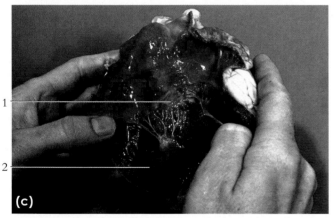

(c)

Continue the incision through the right ventricle to the apex of the heart. Observe the structure of the valve.
1. Right atrioventricular valve
2. Right ventricle

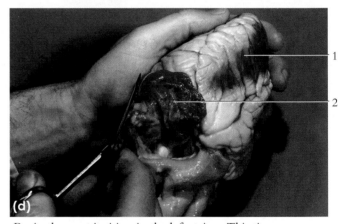

(d)

Begin the next incision in the left atrium. This time, continue through both the atrium and the ventricle.
1. Left ventricle
2. Left atrium

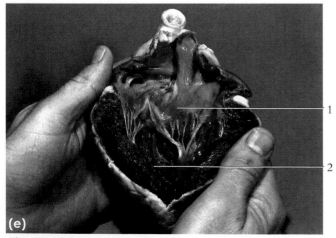

(e)

Expose the left ventricle and atrium. Notice the difference between the right and left ventricles, especially the thicker muscular wall of the left ventricle.
1. Left atrioventricular (bicuspid) valve
2. Left ventricle

Figure 12.38 The steps of a sheep heart dissection (a–e).

Lymphatic System

The lymphatic system is closely interrelated to the circulatory system. The functions of the lymphatic system are basically fourfold: 1) it transports excess interstitial (tissue) fluid, which was initially formed as a blood filtrate, back to the bloodstream; 2) it maintains homeostasis around body cells by providing a constantly moist intercellular environment, which assists movements of materials into and out of cells; 3) it serves as the route by which absorbed fat from the small intestine is transported to the blood; and 4) it helps provide immunological defenses against disease-causing agents.

Lymph capillaries drain tissue fluid, which is formed from blood plasma; when this fluid enters lymph capillaries, it is called **lymph**. Lymph is returned to the venous system via two large lymph ducts—the **thoracic duct** and the **right lymphatic duct** (Fig. 13.1). On the way to these drainage ducts, lymph filters through **lymph nodes**, which contain phagocytic cells and germinal centers containing lymphocytes (made in bone marrow). The **spleen** and **thymus** are considered lymphoid organs because they also produce mature lymphocytes.

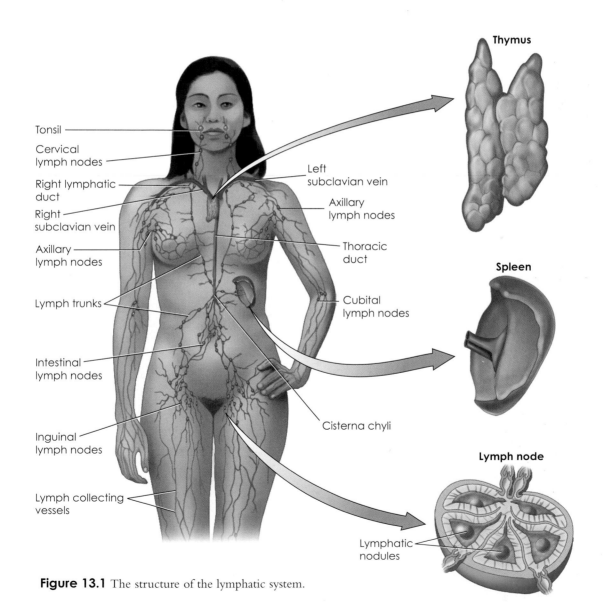

Figure 13.1 The structure of the lymphatic system.

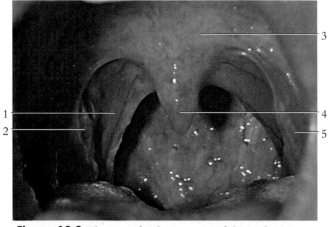

Figure 13.2 The superficial structures of the oral cavity.
1. Pharyngopalatine arch
2. Palatine tonsil
3. Soft palate
4. Palatine uvula
5. Glossopalatine arch

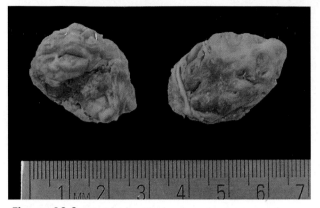

Figure 13.3 Palatine tonsils that have been removed in a tonsillectomy. Chronic tonsillitis generally requires a tonsillectomy.

Figure 13.4 Palatine tonsil.
1. Oral mucosa
2. Lymphatic nodule
3. Germinal centers

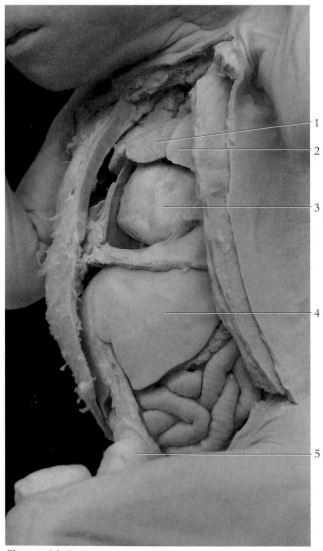

Figure 13.5 The thymus within a fetus, during the third trimester of development.
1. Thymus
2. Lung
3. Heart
4. Liver
5. Umbilical cord

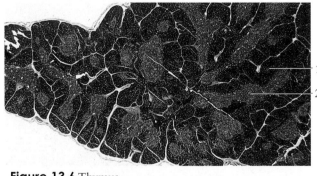

Figure 13.6 Thymus.
1. Cortex
2. Medulla

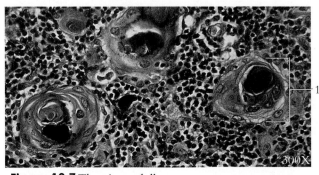

Figure 13.7 Thymic medulla.
1. Hassall's corpuscle

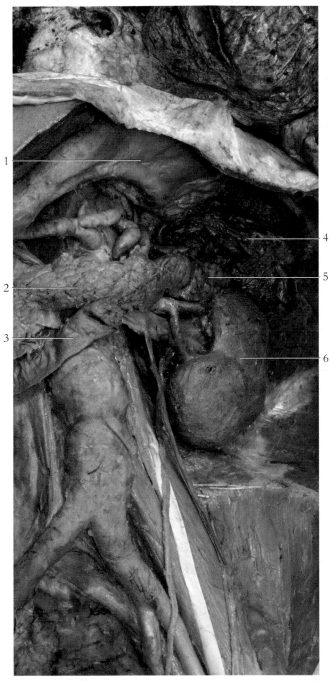

Figure 13.8 The spleen within the abdominal cavity with overlying viscera removed.

1. Stomach (reflected)   4. Spleen
2. Pancreas   5. Adrenal gland
3. Duodenum   6. Kidney

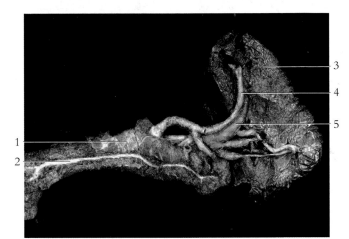

Figure 13.9 The spleen and pancreas.
1. Pancreas
2. Pancreatic duct
3. Spleen
4. Splenic artery
5. Splenic vein

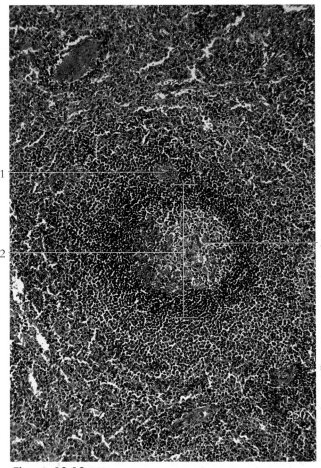

Figure 13.10 Spleen.
1. Central artery
2. Splenic nodule
3. Germinal center

Figure 13.11 A lymph node.

1. Lymph node	3. Lymphatic vessels
2. Vein	4. Muscle

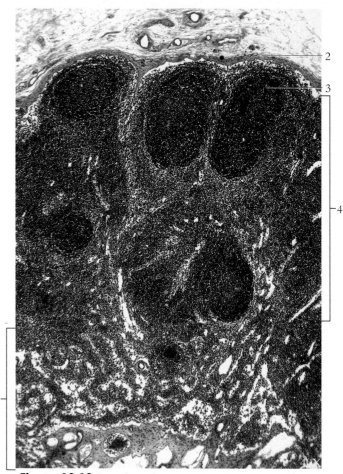

Figure 13.12 Lymph node.

1. Medulla of lymph node	3. Lymphatic nodule
2. Capsule	4. Cortex of lymph node

Respiratory System

The respiratory system is made up of organs and structures that atmospheric oxygen to diffuse into blood, and blood carbon dioxide to diffuse out of blood into air. This system consists of the **nasal cavity, pharynx, larynx,** and **trachea,** and the **bronchi, bronchioles,** and **pulmonary alveoli** within the lungs (Fig. 14.1). The functions of the respiratory system are gas exchange, sound production, assistance in abdominal compression, and coughing and sneezing.

The nasal cavity has a bony and cartilaginous support. The ciliated, mucous lining of the upper respiratory tract warms, moistens, and cleanses inspired air. The **paranasal sinuses** are found in the maxillary, frontal, sphenoid, and ethmoid bones. The **pharynx** is a funnel-shaped passageway that connects the oral and nasal cavities with the larynx. The cartilaginous **larynx** keeps the passageway to the trachea open during breathing and closes the respiratory passageway during swallowing. It also contains the **vocal folds (vocal cords).** The **trachea** is a rigid tube, supported by C-rings of cartilage, that leads from the larynx to the **bronchial tree. Pulmonary alveoli** are the functional units of the lungs where gas exchange occurs; they are small, numerous, thin-walled air sacs. The right and left **lungs** are separated by the **mediastinum.** Each lung is divided into **lobes** and **lobules** and is contained within a pleural cavity lined by the visceral and parietal pleura. The right lung has three lobes and the left lung has two lobes.

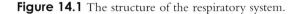

Figure 14.1 The structure of the respiratory system.

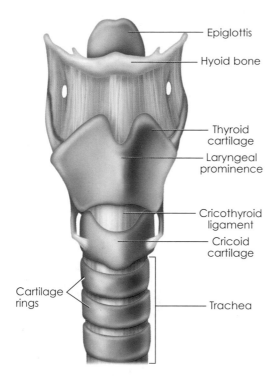

- Epiglottis
- Hyoid bone
- Thyroid cartilage
- Laryngeal prominence
- Cricothyroid ligament
- Cricoid cartilage
- Cartilage rings
- Trachea

Figure 14.2 An anterior view of the larynx and trachea.

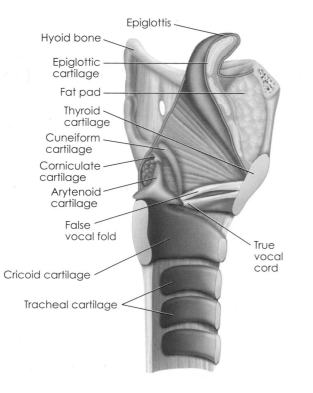

- Epiglottis
- Hyoid bone
- Epiglottic cartilage
- Fat pad
- Thyroid cartilage
- Cuneiform cartilage
- Corniculate cartilage
- Arytenoid cartilage
- False vocal fold
- Cricoid cartilage
- Tracheal cartilage
- True vocal cord

Figure 14.3 A lateral view of the larynx and trachea.

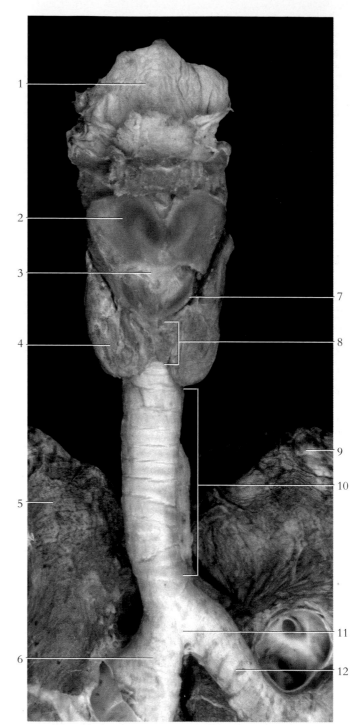

Figure 14.4 An anterior view of the larynx and trachea.

1. Epiglottis
2. Thyroid cartilage
3. Cricoid cartilage
4. Thyroid gland
5. Superior lobe of right lung
6. Right principal (primary) bronchus
7. Cricothyroid ligament
8. Isthmus of thyroid gland
9. Superior lobe of left lung
10. Trachea
11. Carina
12. Left principal (primary) bronchus

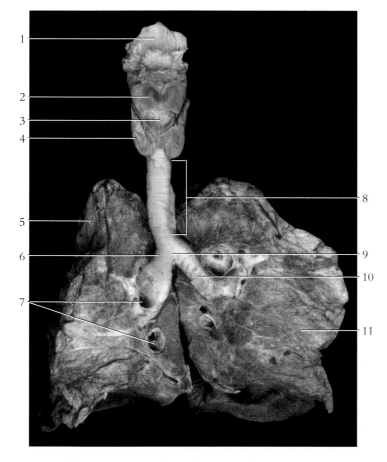

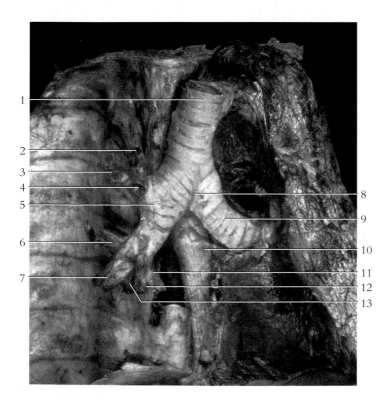

Figure 14.5 An anterior view of the larynx, trachea, and lungs.

1. Epiglottis
2. Thyroid cartilage
3. Cricoid cartilage
4. Thyroid gland
5. Right lung
6. Right principal (primary) bronchus
7. Pulmonary vessels
8. Trachea
9. Carina
10. Left principal (primary) bronchus
11. Left lung

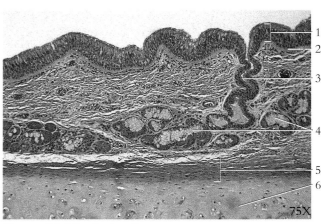

Figure 14.6 Tracheal wall.

1. Respiratory epithelium
2. Basement membrane
3. Duct of seromucous gland
4. Seromucous glands
5. Perichondrium
6. Hyaline cartilage

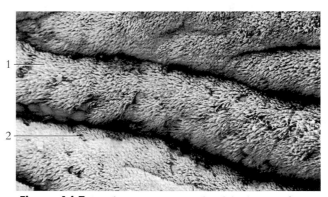

Figure 14.7 An electron micrograph of the lining of the trachea.

1. Cilia
2. Goblet cell

Figure 14.8 An anterior view of bronchi.

1. Trachea
2. Apical segmental bronchus
3. Posterior segmental bronchus
4. Anterior segmental bronchus
5. Right principal bronchus
6. Medial segmental bronchus
7. Anterior basal segmental bronchus
8. Carina
9. Left principal bronchus
10. Esophagus
11. Medial basal segmental bronchus
12. Posterior basal segmental bronchus
13. Lateral basal segmental bronchus

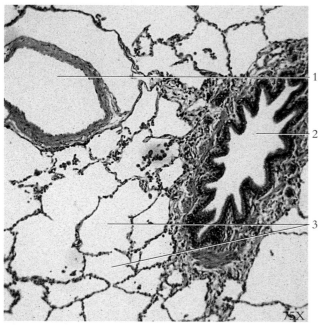

75X

Figure 14.9 Bronchiole.
1. Pulmonary arteriole
2. Bronchiole
3. Pulmonary alveoli

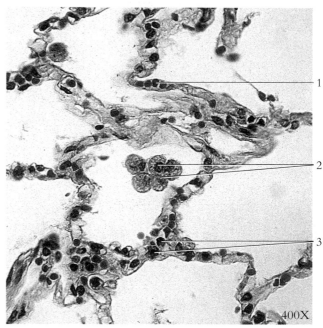

400X

Figure 14.10 Pulmonary alveoli.
1. Capillary in alveolar wall   3. Type II pneumocytes
2. Macrophages

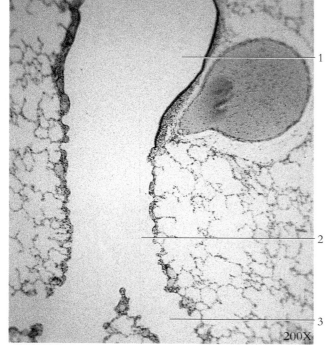

200X

Figure 14.11 Terminal bronchiole.
1. Terminal bronchiole   3. Alveolar duct
2. Respiratory bronchiole

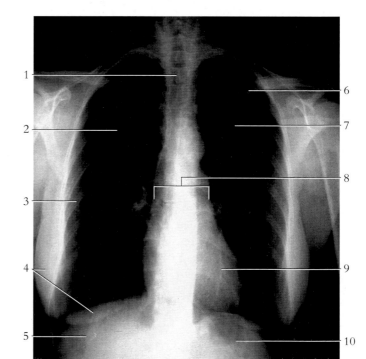

Figure 14.12 A radiograph of the thorax.
1. Thoracic vertebra   6. Clavicle
2. Right lung   7. Left lung
3. Rib   8. Mediastinum
4. Image of right breast   9. Heart
5. Diaphragm/liver   10. Diaphragm/stomach

Digestive System

Chapter 15

The digestive system consists of a **gastrointestinal tract (GI tract)** and **accessory digestive organs**. Most of the food we eat is not suitable for cellular utilization until it is mechanically and chemically reduced to forms that can be absorbed through the intestinal wall and transported to the cells by the blood or lymph. Ingested food is not technically in the body until it is absorbed and, in fact, a large portion of consumed food is not digested at all but rather passes through as fecal material.

The functions of the principal regions and organs of the digestive system are presented in Table 15.1. The digestive system is diagrammed in Figure 15.1.

Table 15.1 Regions and structures of the digestive system.

Region or Structure	Function
Gastrointestinal tract	
Oral cavity	Ingests food; receives saliva and initiates digestion of carbohydrates; mastication (chewing); forms bolus (food mass); deglutition (swallowing)
Pharynx	Receives bolus from oral cavity and passes it to esophagus
Esophagus	Transports bolus to stomach by peristalsis
Stomach	Receives bolus from esophagus; forms chyme (paste-like food) initiates digestion of proteins; moves chyme into duodenum; participates in vomiting
Small intestine	Receives chyme from stomach, along with secretions from liver and pancreas; chemically and mechanically breaks down chyme; absorbs nutrients; transports wastes to large intestine
Large intestine	Receives undigested wastes from small intestine; absorbs water and electrolytes; provides for further bacterial digestion, forms and stores feces, and expels feces through defecation
Accessory digestive organs	
Teeth	Mechanically pulverize food
Tongue	Manipulates food and assists in swallowing
Salivary glands	Secrete saliva which aids in formation of bolus; initiates digestion of carbohydrates
Liver	Production of bile; storage of iron and copper; interconversion of glucose to glycogen and storage of glycogen; synthesis of certain vitamins; production of urea; synthesis of fibrinogen and prothrombin used for clotting of blood; phagocytosis of foreign material in blood; detoxifies harmful substances in body; storage of blood cells; hemopoiesis in fetus and newborn
Gallbladder	Concentration and storage of bile necessary for emulsification of fats; releases bile into duodenum
Pancreas	Production and secretion of pancreatic juice containing digestive enzymes; production and secretion of the hormones insulin and glucagon

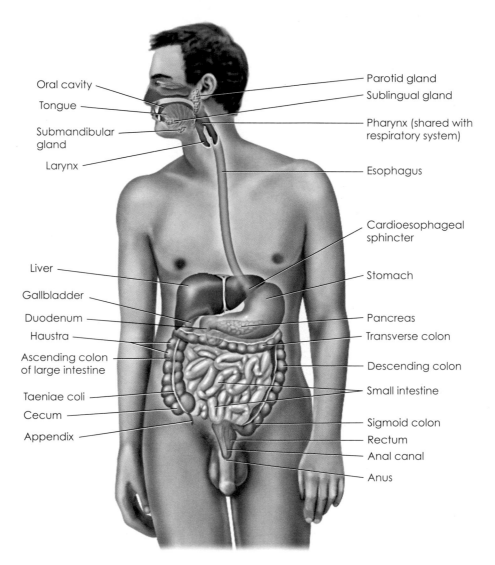

Figure 15.1 The structure of the digestive system.

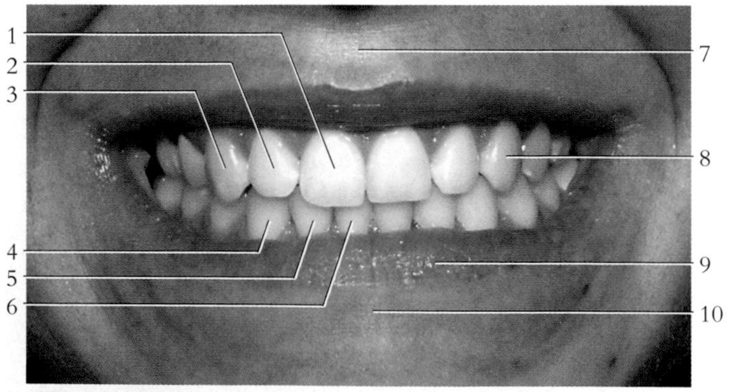

Oral cavity
Tongue
Submandibular gland
Larynx

Parotid gland
Sublingual gland
Pharynx (shared with respiratory system)
Esophagus

Cardioesophageal sphincter

Liver
Gallbladder
Duodenum
Haustra
Ascending colon of large intestine
Taeniae coli
Cecum
Appendix

Stomach

Pancreas
Transverse colon
Descending colon
Small intestine
Sigmoid colon
Rectum
Anal canal
Anus

Figure 15.2 The oral region, lips, and teeth.

1. Medial (central) incisor
2. Lateral incisor
3. Canine
4. Canine
5. Lateral incisor
6. Medial (central) incisor
7. Philtrum
8. Canine
9. Inferior lip
10. Mentolabial sulcus

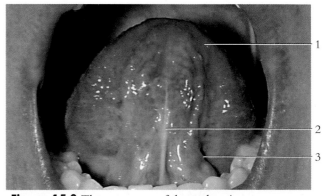

Figure 15.3 The structures of the oral cavity with the mouth open and the tongue elevated.

1. Tongue
2. Lingual frenulum
3. Opening of submandibular duct

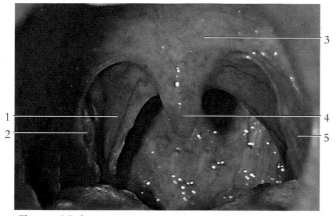

Figure 15.4 The superficial structures of the oral cavity.
1. Pharyngopalatine arch   4. Palatine uvula
2. Palatine tonsil   5. Glossopalatine arch
3. Soft palate

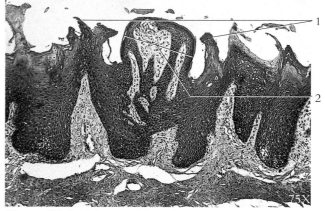

Figure 15.5 Filiform and fungiform papillac.
1. Filiform papillae
2. Fungiform papilla

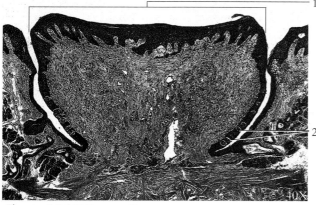

Figure 15.6 Vallate papilla
1. Vallate papilla
2. Taste buds

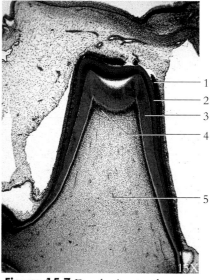

Figure 15.7 Developing tooth.
1. Ameloblasts   3. Dentin   5. Pulp
2. Enamel   4. Odontoblasts

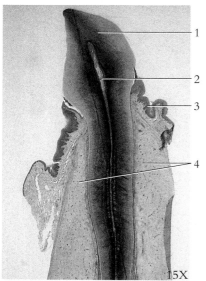

Figure 15.8 Mature tooth.
1. Dentin (enamel has   3. Gingiva
   been dissolved away)   4. Alveolar bone
2. Pulp

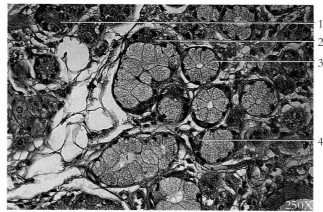

Figure 15.9 Acini of salivary tissue.
1. Serous acinus   3. Mucous acinus
2. Serous demilune on   4. Serous demilune on
   mucous acinus   mucous acinus

Figure 15.10 Striated duct of salivary gland.
1. Mucous acini
2. Serous acinus
3. Lumen of striated duct

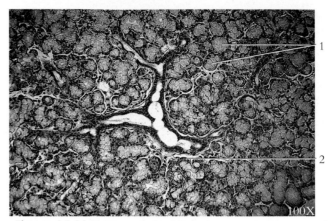

Figure 15.11 Sublingual gland (mostly mucous, some serous).
1. Mucous acini
2. Serous demilune

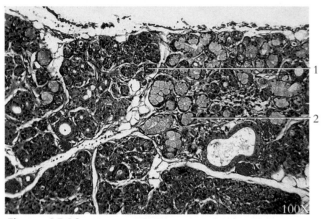

Figure 15.12 Submandibular gland (about 1/5 mucous and 4/5 serous).
1. Serous acinus
2. Mucous acinus

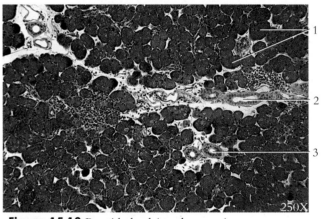

Figure 15.13 Parotid gland (purely serous).
1. Serous acini
2. Lumen of excretory duct
3. Lumen of striated duct

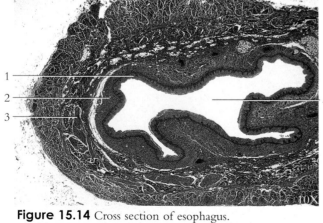

Figure 15.14 Cross section of esophagus.
1. Mucosa
2. Submucosa
3. Muscularis externa
4. Lumen

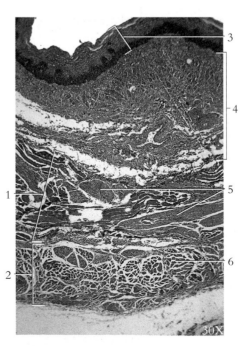

Figure 15.15 Wall of esophagus.
1. Inner circular layer (muscularis externa)
2. Outer longitudinal layer (muscularis externa)
3. Mucosa
4. Submucosa
5. Smooth muscle
6. Skeletal muscle

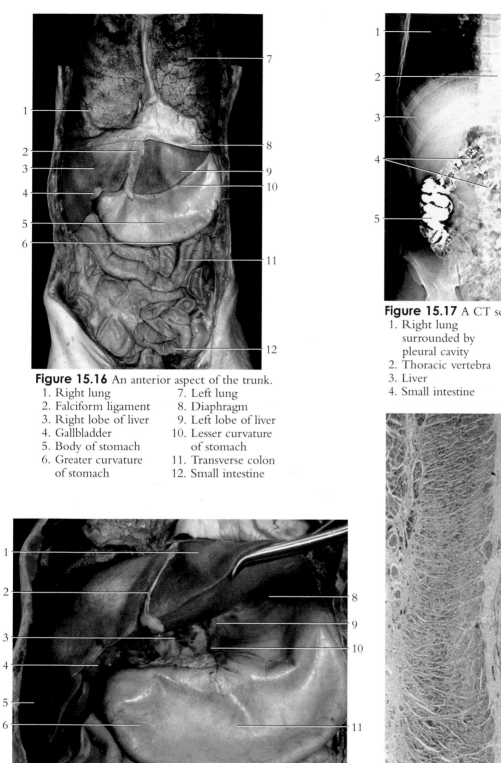

Figure 15.16 An anterior aspect of the trunk.

1. Right lung
2. Falciform ligament
3. Right lobe of liver
4. Gallbladder
5. Body of stomach
6. Greater curvature of stomach
7. Left lung
8. Diaphragm
9. Left lobe of liver
10. Lesser curvature of stomach
11. Transverse colon
12. Small intestine

Figure 15.17 A CT scanogram of the trunk.

1. Right lung surrounded by pleural cavity
2. Thoracic vertebra
3. Liver
4. Small intestine
5. Ascending colon
6. Heart
7. Diaphragm
8. Stomach
9. Descending colon
10. Ilium

Figure 15.18 An anterior view of the stomach and liver.

1. Left lobe of liver (reflected)
2. Falciform ligament
3. Celiac trunk (covered by hepatoduodenal ligament)
4. Gallbladder
5. Right lobe of liver
6. Pylorus of stomach
7. Transverse colon
8. Fundus of stomach
9. Lesser omentum
10. Gastric vein (traversing through omentum)
11. Body of stomach

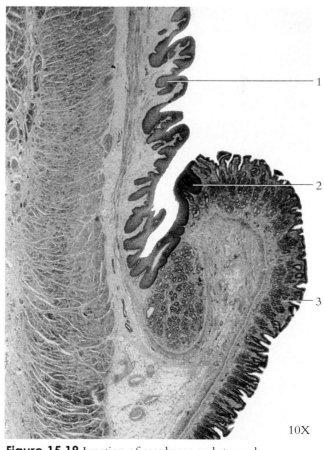

10X

Figure 15.19 Junction of esophagus and stomach.

1. Epithelium of esophagus
2. Abrupt change in epithelium
3. Epithelium of stomach

143

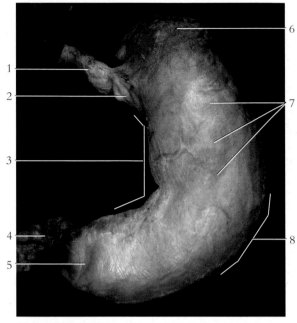

Figure 15.20 The major regions and structures of the stomach.

1. Esophagus
2. Cardiac portion of stomach
3. Lesser curvature of stomach
4. Duodenum
5. Pylorus of stomach
6. Fundus of stomach
7. Body of stomach
8. Greater curvature of stomach

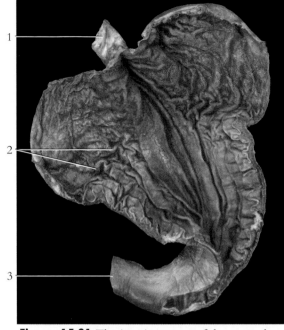

Figure 15.21 The interior aspect of the stomach.

1. Esophagus
2. Gastric folds (rugae)
3. Duodenum

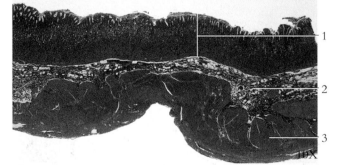

Figure 15.22 Wall of stomach.

1. Mucosa
2. Submucosa
3. Muscularis externa

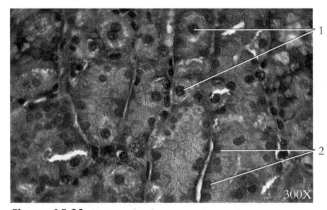

Figure 15.23 Gastric glands.

1. Parietal cells
2. Chief cells

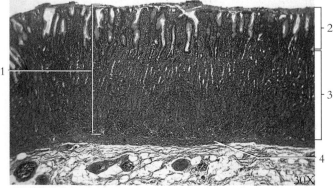

Figure 15.24 Mucosa of stomach body.

1. Mucosa
2. Gastric pits
3. Gastric glands
4. Muscularis mucosae

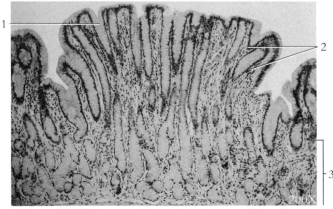

Figure 15.25 Mucosa of stomach pylorus.

1. Gastric pit
2. Lamina propria
3. Gastric glands

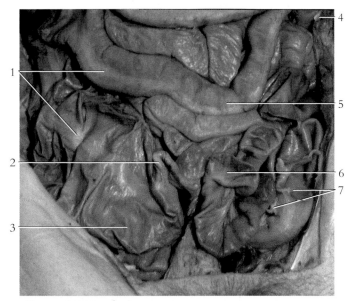

Figure 15.26 An anterior aspect of the small and large intestines with associated structures.

1. Taeniae coli
2. Appendix
3. Cecum
4. Left colic (splenic) flexure
5. Transverse colon
6. Small intestine
7. Epiploic appendages

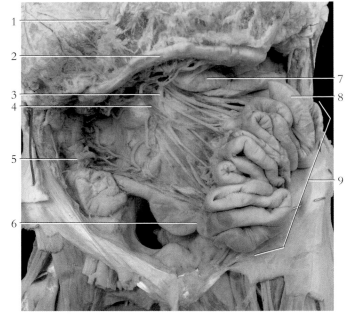

Figure 15.27 An anterior aspect of the small and large intestines with associated structures.

1. Greater omentum (reflected)
2. Transverse colon (reflected)
3. Superior mesenteric artery
4. Superior mesenteric vein
5. Ascending colon
6. Ileum
7. Stomach
8. Jejunum
9. Small intestine (reflected)

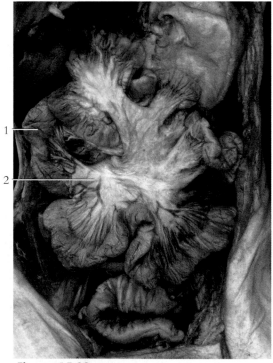

Figure 15.28 The mesentery and small intestine within abdominal cavity.

1. Serosa (visceral peritoneum)
2. Mesentery

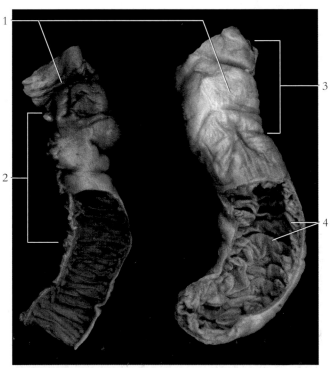

Figure 15.29 Sections of the small intestine.

1. Serosa (visceral peritoneum)
2. Ileum
3. Jejunum
4. Plicae circulares

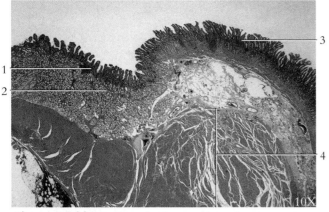

Figure 15.30 Junction of pylorus and duodenum.
1. Duodenal epithelium
2. Duodenal (Brunner's) glands
3. Stomach epithelium
4. Pyloric sphincter

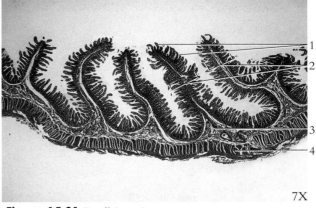

Figure 15.31 Small intestine.
1. Intestinal villi
2. Plicae circulares
3. Submucosa
4. Muscularis externa

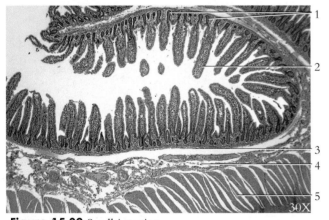

Figure 15.32 Small intestine.
1. Intestinal glands
2. Intestinal villi
3. Muscularis mucosae
4. Submucosa
5. Muscularis externa

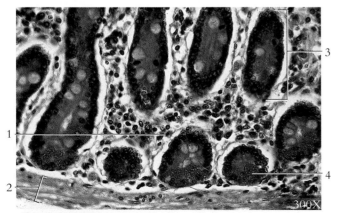

Figure 15.33 Intestinal glands.
1. Eosinophil within lamina propria
2. Muscularis mucosae
3. Intestinal gland
4. Paneth cells

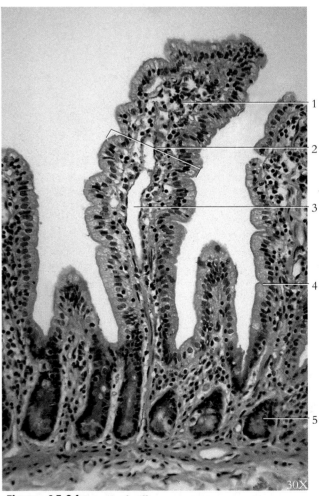

Figure 15.34 Intestinal villus.
1. Lamina propria
2. Intestinal villus
3. Central lacteal
4. Mucosa (simple columnar epithelium)
5. Intestinal gland

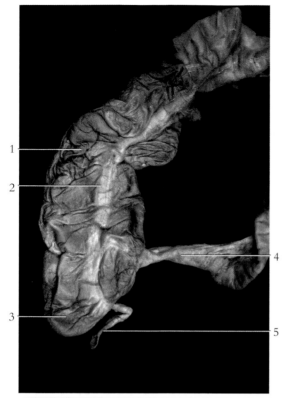

Figure 15.35 The cecum and appendix.
1. Ascending colon   4. Ileum
2. Taeniae coli   5. Appendix
3. Cecum

Figure 15.36 A radiograph of the large intestine.
1. Right colon (hepatic) flexure   7. Transverse colon
2. Ascending colon   8. Lumbar vertebra
3. Sigmoid colon   9. Descending colon
4. Cecum   10. Rectum
5. Left colic (splenic) flexure   11. Hip joint
6. 12th rib

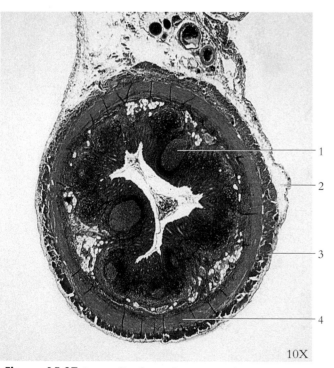

10X

Figure 15.37 Appendix, shown in cross section.
1. Lymphatic nodule
2. Serosa (visceral peritoneum)
3. Longitudinal layer of muscularis externa
4. Circular layer of muscularis externa

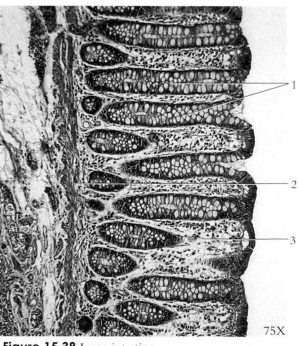

75X

Figure 15.38 Large intestine.
1. Glands
2. Muscularis mucosae
3. Lamina propria

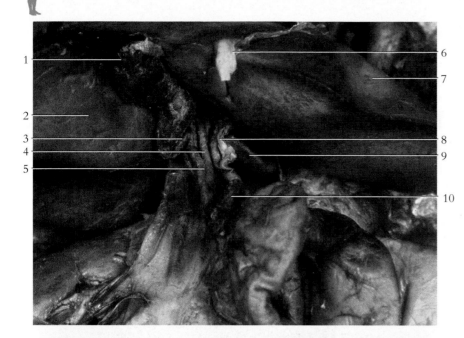

Figure 15.39 An inferior view of the liver and gallbladder.
1. Gallbladder
2. Right lobe of liver
3. Cystic duct
4. Hepatic duct
5. Common bile duct
6. Falciform ligament
7. Left lobe of liver
8. Hepatic artery
9. Caudate lobe of liver
10. Hepatic portal vein

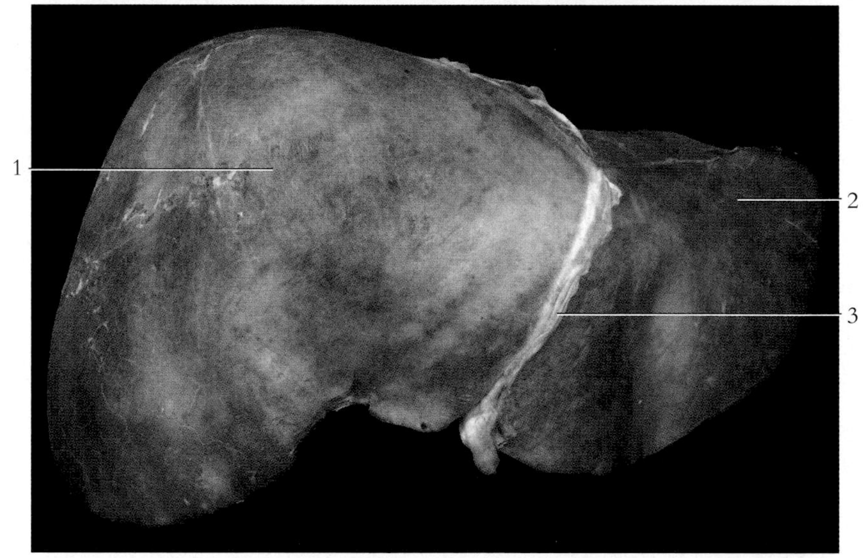

Figure 15.40 An anterior view of the liver.
1. Right lobe of liver
2. Left lobe of liver
3. Falciform ligament

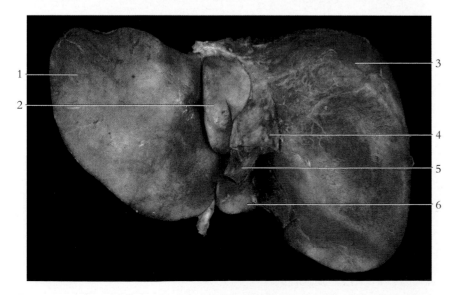

Figure 15.41 An inferior view of the liver.
1. Left lobe of liver
2. Caudate lobe of liver
3. Right lobe of liver
4. Inferior vena cava
5. Hepatic portal vein
6. Quadrate lobe of liver

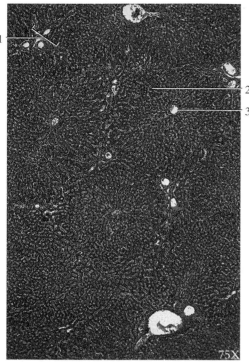

Figure 15.42 Hepatic lobules.
1. Portal area   3. Central vein
2. Hepatic lobule

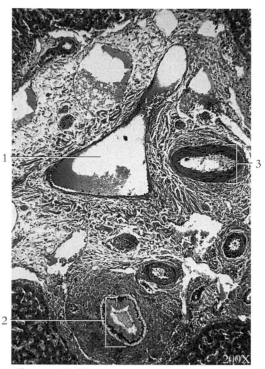

Figure 15.43 Portal area.
1. Lumen of   2. Bile duct
   portal vein   3. Hepatic artery

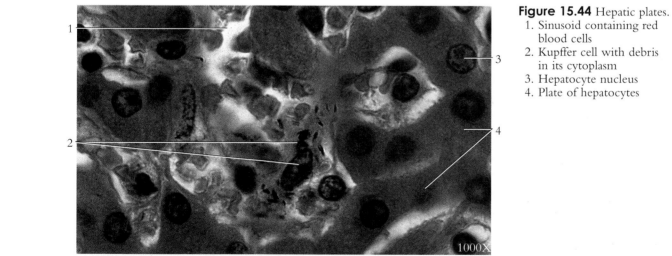

Figure 15.44 Hepatic plates.
1. Sinusoid containing red blood cells
2. Kupffer cell with debris in its cytoplasm
3. Hepatocyte nucleus
4. Plate of hepatocytes

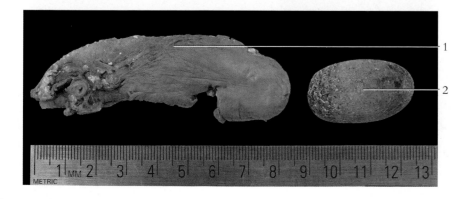

Figure 15.45 A gallbladder that has been removed (cholecystectomy) and cut open to remove its gallstone (biliary calculus).
1. Gallbladder
2. Gallstone

Figure 15.46 Gallbladder.
1. Mucosal folds   3. Muscularis
2. Lamina propria

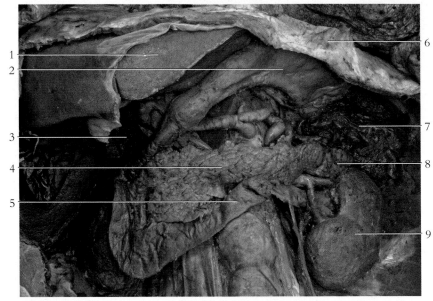

Figure 15.47 An anterior view of abdominal cavity showing pancreas and associated structures with overlying viscera removed.
1. Liver (reflected)
2. Stomach (reflected)
3. Gallbladder
4. Pancreas
5. Duodenum
6. Diaphragm
7. Spleen
8. Adrenal gland
9. Kidney

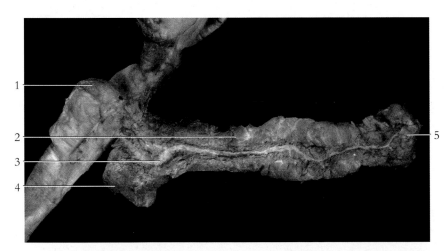

Figure 15.48 An anterior view of the pancreas and pancreatic duct.
1. Duodenum
2. Body of pancreas
3. Pancreatic duct
4. Head of pancreas
5. Tail of pancreas

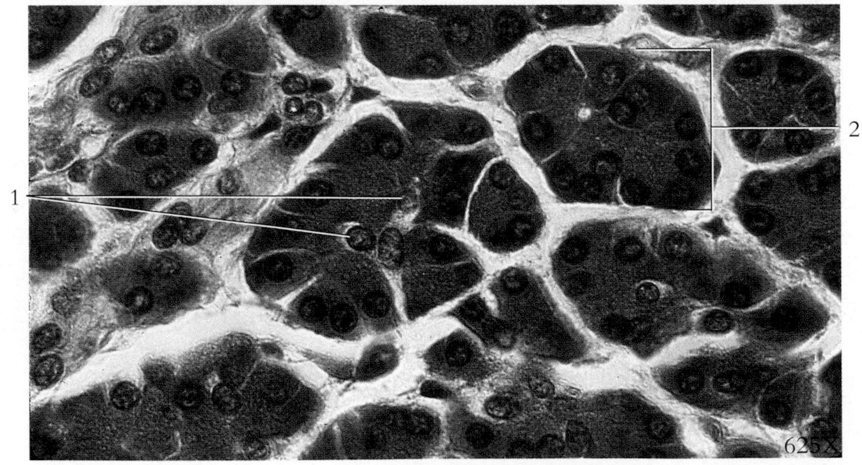

Figure 15.49 Pancreatic acini.
1. Centroacinar cells
2. Acinus

Urinary System

The urinary system consists of the **kidneys, ureters, urinary bladder**, and **urethra** (Fig. 16.1). The urinary system: 1) removes metabolic wastes from the blood and excretes it (as urine) to the outside during micturition (the physiological aspect of urination); 2) regulates, in part, the rate of red blood cell formation by secretion of the hormone **erythropoietin;** 3) assists the regulation of blood pressure by secreting the enzyme **renin;** 4) assists the regulation of calcium homeostasis by activating vitamin D; and 5) assists the regulation of the volume, composition, and pH of body fluids.

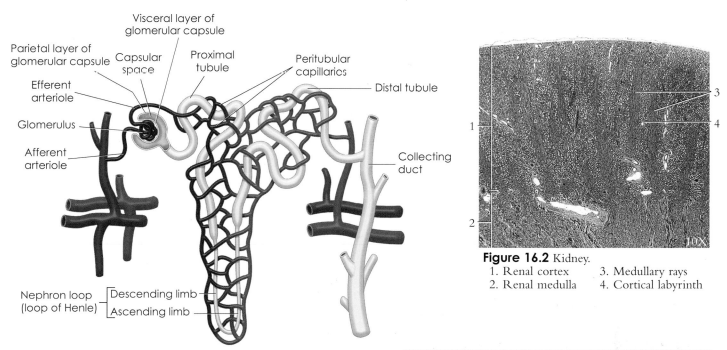

Figure 16.1 The structure of the nephron.

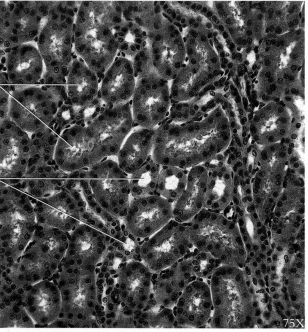

Figure 16.2 Kidney.
1. Renal cortex
2. Renal medulla
3. Medullary rays
4. Cortical labyrinth

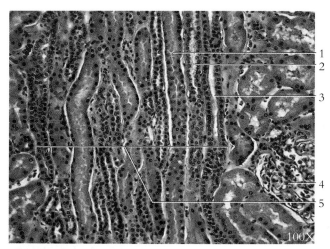

Figure 16.3 Medullary ray.
1. Proximal tubule (of straight portion)
2. Distal tubule (of straight portion)
3. Collecting tubule
4. Glomerulus
5. Medullary ray

Figure 16.4 Cortical labyrinth.
1. Proximal convoluted tubule
2. Distal convoluted tubule

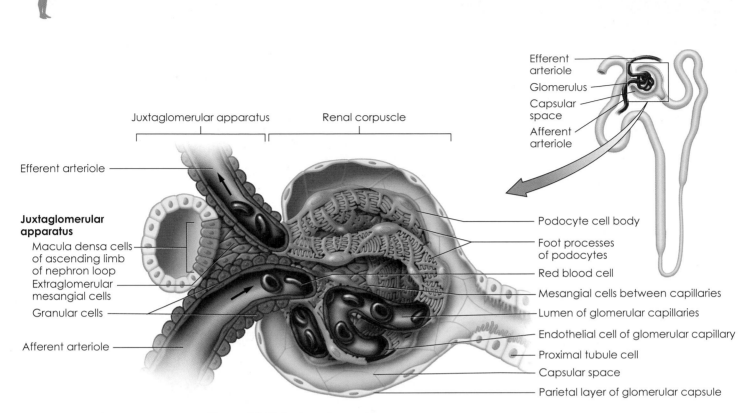

Juxtaglomerular apparatus

Renal corpuscle

Efferent arteriole

Glomerulus

Capsular space

Afferent arteriole

Efferent arteriole

Juxtaglomerular apparatus

Macula densa cells of ascending limb of nephron loop

Extraglomerular mesangial cells

Granular cells

Afferent arteriole

Podocyte cell body

Foot processes of podocytes

Red blood cell

Mesangial cells between capillaries

Lumen of glomerular capillaries

Endothelial cell of glomerular capillary

Proximal tubule cell

Capsular space

Parietal layer of glomerular capsule

Figure 16.5 The juxtaglomerular apparatus.

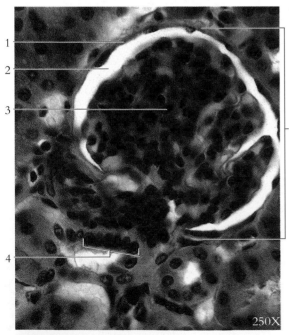

Figure 16.6 A renal corpuscle.
1. Glomerular capsule
2. Capsular space
3. Glomerulus
4. Macula densa of distal tubule
5. Renal corpuscle

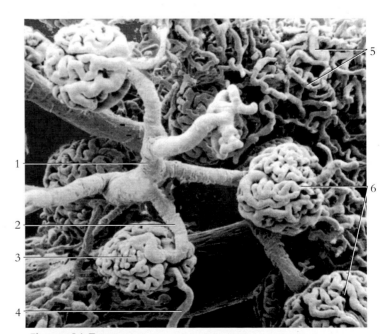

Figure 16.7 An electron micrograph of the renal cortex.
1. Interlobular artery
2. Afferent glomerular arteriole
3. Glomerulus
4. Efferent glomerular arteriole
5. Peritubular capillaries
6. Glomeruli

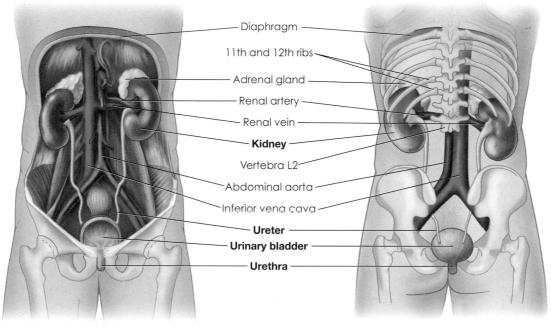

Figure 16.8 The organs of the urinary system.

Anterior view

Posterior view

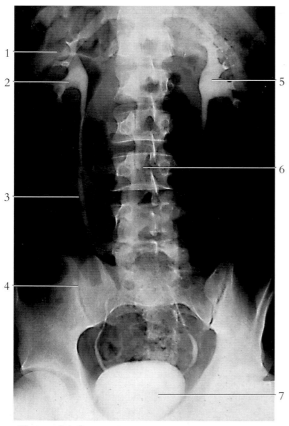

Figure 16.9 A pyelogram of the urinary system.
1. 12th rib   5. Renal pelvis
2. Major calyx   6. 3rd lumbar vertebra
3. Ureter   7. Urinary bladder
4. Sacroiliac joint

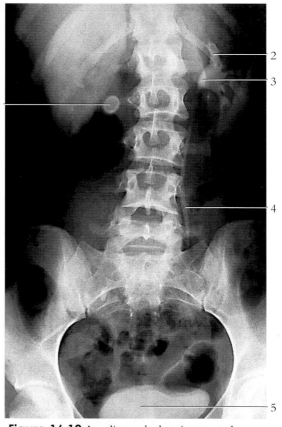

Figure 16.10 A radiograph showing a renal calculus (kidney stone) in the renal pelvis of the right kidney.
1. Renal calculus   4. Left ureter
2. Major calyx of left kidney   5. Urinary bladder
3. Renal pelvis of kidney

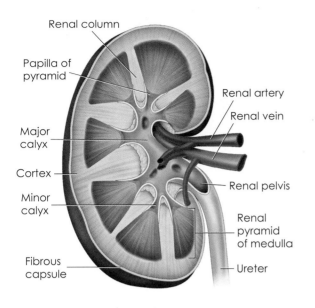

Renal column

Papilla of pyramid

Major calyx

Cortex

Minor calyx

Fibrous capsule

Renal artery

Renal vein

Renal pelvis

Renal pyramid of medulla

Ureter

Figure 16.11 The structure of the kidney.

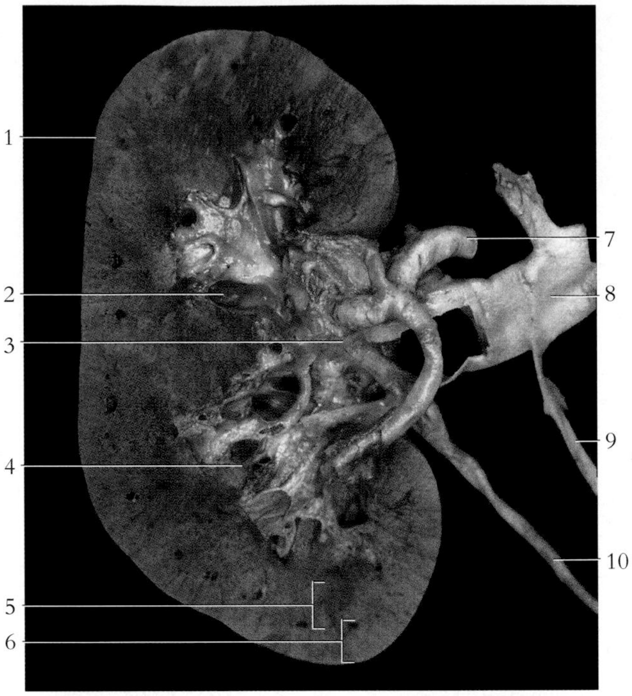

Figure 16.12 A coronal section of the left kidney.
1. Renal capsule
2. Major calyx
3. Renal pelvis
4. Renal papilla
5. Renal medulla
6. Renal cortex
7. Renal artery
8. Renal vein
9. Left testicular vein
10. Ureter

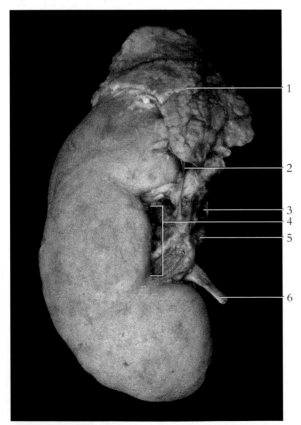

Figure 16.13 An anterior view of the right kidney.
1. Adrenal gland
2. Suprarenal artery
3. Renal artery
4. Hilum
5. Renal vein
6. Ureter

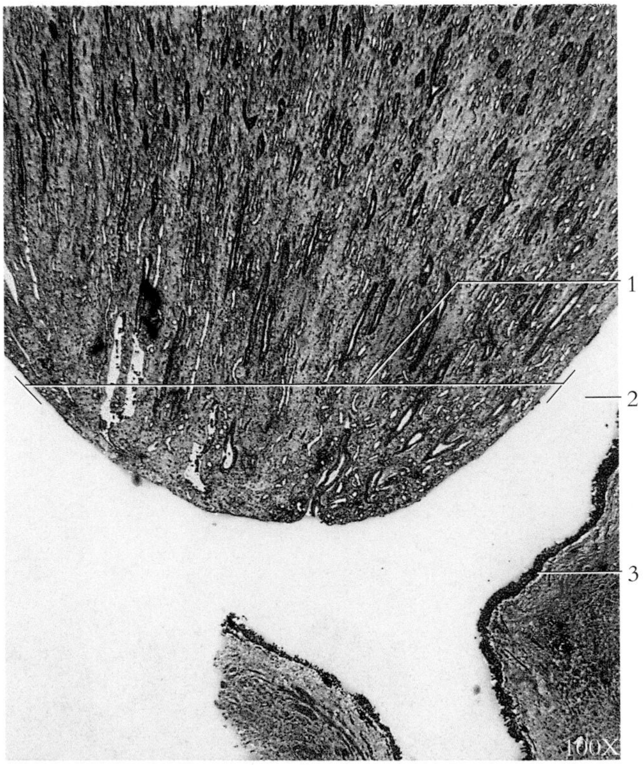

Figure 16.14 Renal papilla.
1. Renal papilla
2. Minor calyx
3. Transitional epithelium

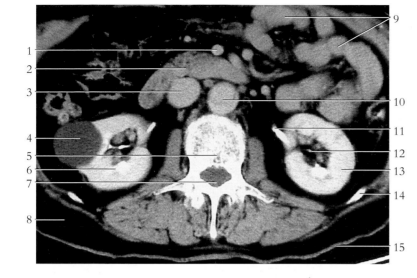

Figure 16.15 A transaxial CT image of abdominal viscera. Note the large renal cyst in the right kidney.
1. Superior mesenteric artery
2. Pancreas
3. Inferior vena cava
4. Renal cyst
5. Body of lumbar vertebra
6. Right kidney
7. Vertebral foramen
8. Subcutaneous fat
9. Small intestine
10. Abdominal portion of aorta
11. Left ureter
12. Renal pelvis
13. Renal cortex
14. Rib
15. Skin

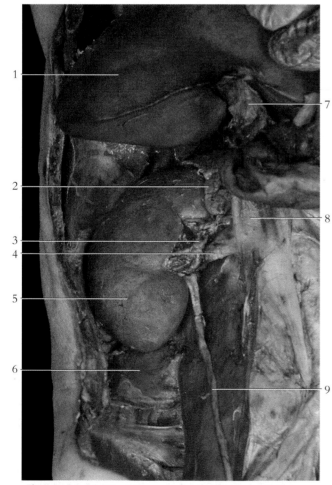

Figure 16.16 The kidney and ureter with overlying viscera removed.
1. Liver
2. Adrenal gland
3. Renal artery
4. Renal vein
5. Right kidney
6. Quadratus lumborum muscle
7. Gallbladder
8. Inferior vena cava
9. Ureter

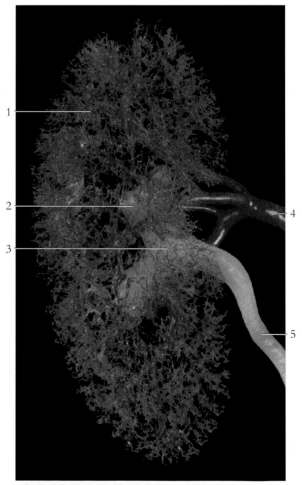

Figure 16.17 A plastic vascular cast of a kidney and ureter.
1. Arteriole network
2. Major calyx
3. Renal pelvis
4. Renal artery
5. Ureter

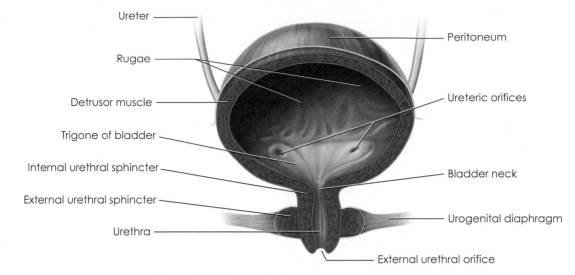

Ureter

Rugae

Detrusor muscle

Trigone of bladder

Internal urethral sphincter

External urethral sphincter

Urethra

Peritoneum

Ureteric orifices

Bladder neck

Urogenital diaphragm

External urethral orifice

Figure 16.18 A frontal view of the female urinary bladder.

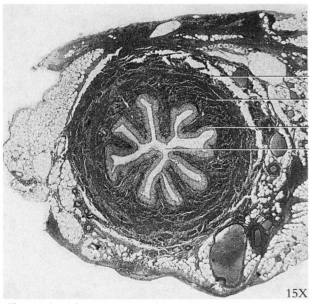

15X

Figure 16.19 Ureter, as shown in cross section.
1. Adventitia   3. Mucosa
2. Muscularis   4. Lumen

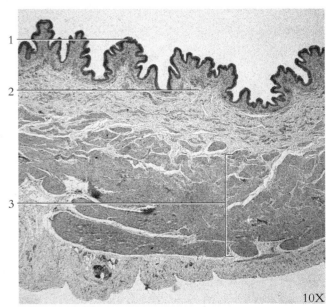

10X

Figure 16.20 Wall of urinary bladder.
1. Transitional epithelium   3. Muscularis
2. Lamina propria

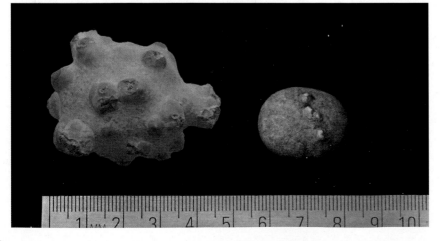

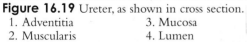

Figure 16.21 Renal calculi (stones) from the urinary bladders of two different patients. Renal calculi vary considerably in appearance and size and may develop in the renal pelvis of the kidney or in the urinary bladder.

Reproductive System

Chapter 17

The organs of the male and female reproductive systems produce gametes and provide the mechanism for the union of these sex cells during the process of coitus (sexual intercourse). It is through sexual reproduction that individuals of a species are propagated, each having a genetic diversity inherited from both parents.

The male reproductive system consists of the **testes** (in the **scrotum**), excretory glands (**seminal vesicles, prostate**, and **bulbourethral glands**), and **penis**. The functions of the male reproductive system are to secrete sex hormones, produce spermatozoa, and ejaculate semen (spermatozoa and additives) into the vagina of the female.

The female reproductive system consists of the **ovaries, uterus, uterine tubes, vagina, external genitalia**, and **mammary glands**. The functions of the female reproductive system are to secrete sex hormones, produce ova, receive ejaculated semen from the erect penis of the male, nourish the developing embryo and fetus, deliver the baby, and nurse the infant once it is born.

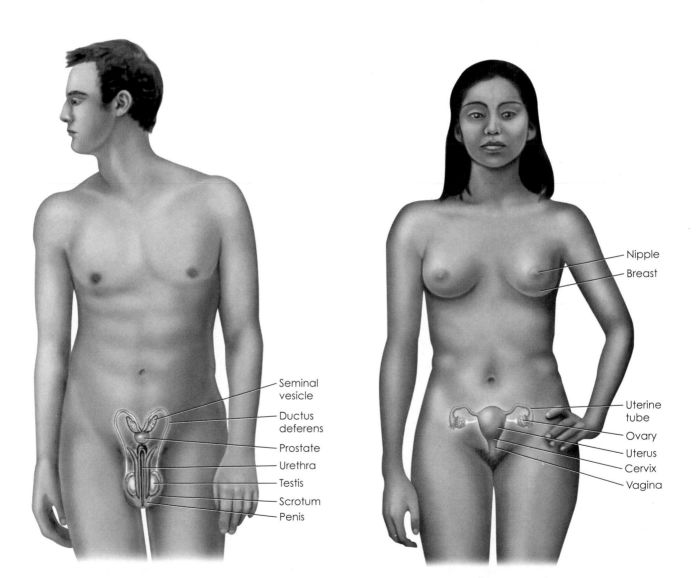

Figure 17.1 The organs of the male reproductive system.

Figure 17.2 The organs of the female reproductive system.

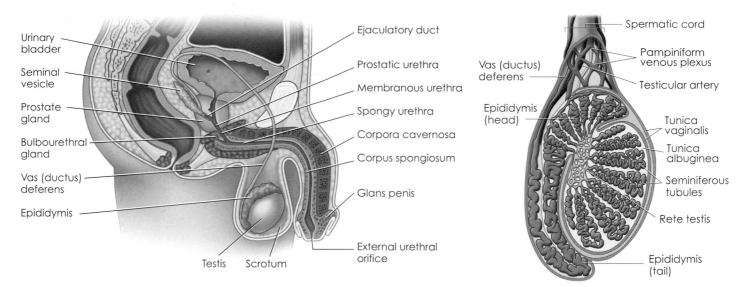

Urinary bladder

Seminal vesicle

Prostate gland

Bulbourethral gland

Vas (ductus) deferens

Epididymis

Ejaculatory duct

Prostatic urethra

Membranous urethra

Spongy urethra

Corpora cavernosa

Corpus spongiosum

Glans penis

External urethral orifice

Testis   Scrotum

Figure 17.3 The structure of the male genitalia.

Spermatic cord

Pampiniform venous plexus

Vas (ductus) deferens

Testicular artery

Epididymis (head)

Tunica vaginalis

Tunica albuginea

Seminiferous tubules

Rete testis

Epididymis (tail)

Figure 17.4 The structure of the testis.

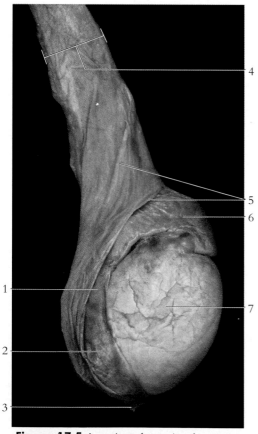

4

5

6

1

7

2

3

Figure 17.5 A testis and associated structures.
1. Body of epididymis   5. Spermatic fascia
2. Tail of epididymis   6. Head of epididymis
3. Gubernaculum   7. Testis
4. Spermatic cord

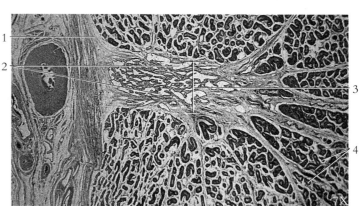

1

2

3

4

Figure 17.6 Testis.
1. Tunic albuginea   3. Mediastinum
2. Tubules of rete testis   4. Seminiferous tubules

Figure 17.7 A cancerous testis; orchiectomy is the removal of a testis.

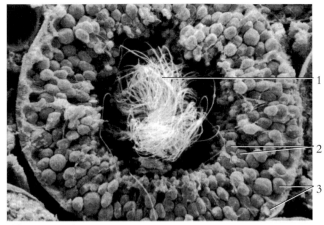
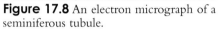

Figure 17.8 An electron micrograph of a seminiferous tubule.
1. Spermatozoa   3. Spermatogonia
2. Spermatids

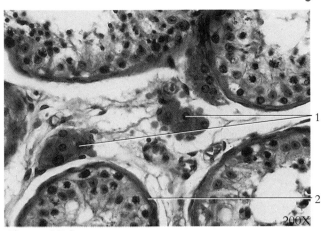

Figure 17.9 Testis.
1. Interstitial (Leydig) cells
2. Seminiferous tubule

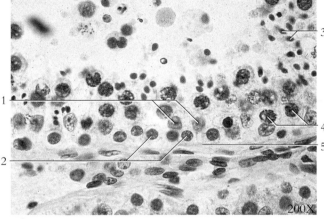

Figure 17.10 Wall of seminiferous tubule.
1. Sustentacular (Sertoli) cells   4. Primary spermatocytes
2. Spermatogonia   5. Boundary of seminiferous
3. Spermatids   tubule

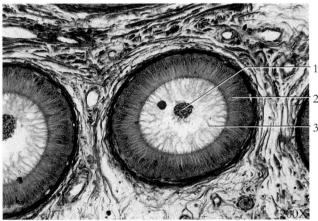

Figure 17.11 Epididymis.
1. Sperm in lumen   3. Stereocilia
2. Pseudostratified columnar
   epithelium

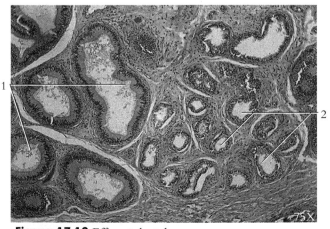

Figure 17.12 Efferent ductules.
1. Duct of epididymis
2. Efferent ductules

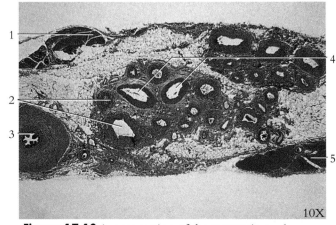

Figure 17.13 A cross section of the spermatic cord.
1. Cremaster muscle   3. Ductus deferens
2. Veins of the pampiniform 4. Testicular arterioles
   plexus   5. Cremaster muscle

159

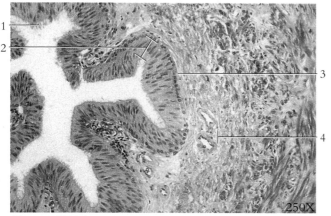

Figure 17.14 Ductus deferens.
1. Stereocilia
2. Pseudostratified columnar epithelium
3. Lamina propria
4. Muscularis

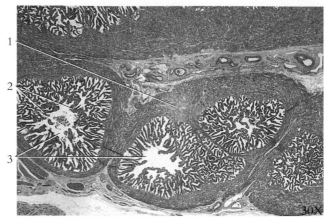

Figure 17.15 Seminal vesicle.
1. Fibromuscular stroma
2. Mucosal folds
3. Lumen

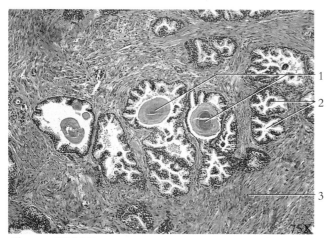

Figure 17.16 Prostate.
1. Prostate concretions
2. Glandular acini
3. Fibromuscular stroma

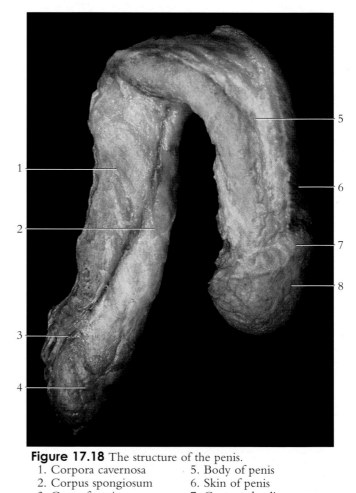

Figure 17.18 The structure of the penis.
1. Corpora cavernosa
2. Corpus spongiosum
3. Crus of penis
4. Bulb of penis
5. Body of penis
6. Skin of penis
7. Corona glandis
8. Glans penis

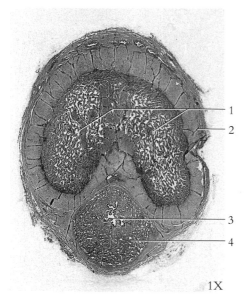

Figure 17.17 Penis.
1. Corpora cavernosa
2. Tunica albuginea
3. Urethra
4. Corpus spongiosum

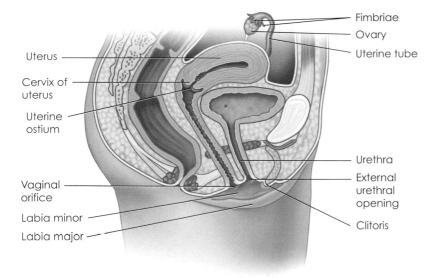

Uterus

Cervix of
uterus

Uterine
ostium

Vaginal
orifice

Labia minor

Labia major

Fimbriae

Ovary

Uterine tube

Urethra

External
urethral
opening

Clitoris

Figure 17.19 The external genitalia and internal
reproductive organs of the female reproductive system.

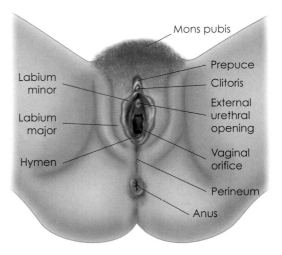

Mons pubis

Labium
minor

Labium
major

Hymen

Prepuce

Clitoris

External
urethral
opening

Vaginal
orifice

Perineum

Anus

Figure 17.20 The external genitalia of the female.

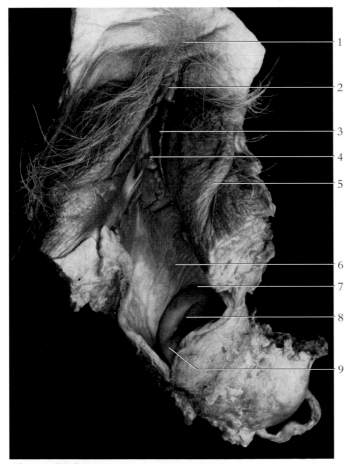

1
2
3
4
5

6
7

8

9

Figure 17.21 The external genitalia and vagina.

1. Mons pubis   6. Vagina (dissected open)
2. Clitoris   7. Fornix
3. Labium minor   8. Uterine ostium
4. Urethral opening   9. Cervix of uterus
5. Labium major

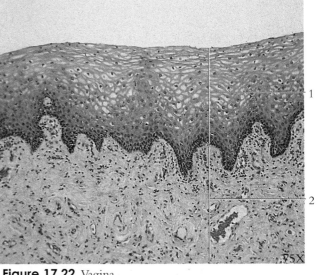

Figure 17.22 Vagina.
1. Non-keratinized stratified squamous epithelium
2. Lamina propria

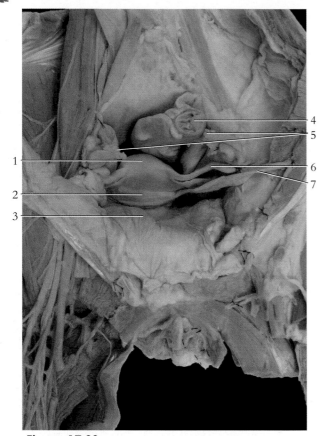

Figure 17.23 The uterus within the pelvic cavity.
1. Fundus of uterus
2. Body of uterus
3. Urinary bladder
4. Sigmoid colon (cut)
5. Fimbriae
6. Uterine tube
7. Suspensory ligament

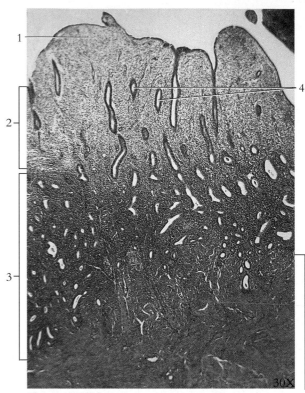

Figure 17.24 Uterus: proliferative phase.
1. Endometrial stroma
2. Lamina functionalis (of endometrium)
3. Lamina basalis (of endometrium)
4. Endometrial uterine glands
5. Myometrium

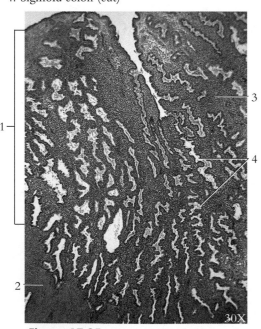

Figure 17.25 Uterus: secretory phase.
1. Endometrium
2. Myometrium
3. Endometrial stroma
4. Endometrial uterine glands

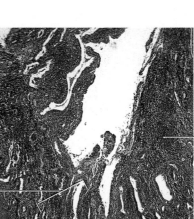

Figure 17.26 Uterus: menstrual phase.
1. Disintegrating lamina functionalis
2. Lamina basalis (still intact)
3. Myometrium
4. Blood in stroma

162

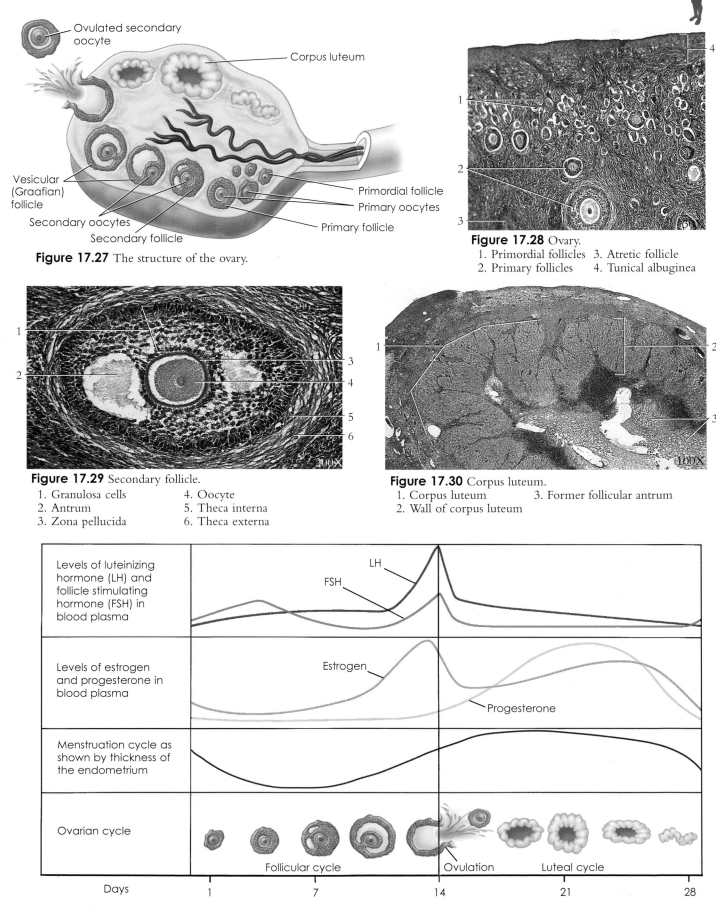

Figure 17.27 The structure of the ovary.

Ovulated secondary oocyte

Corpus luteum

Vesicular (Graafian) follicle

Secondary oocytes

Secondary follicle

Primordial follicle

Primary oocytes

Primary follicle

Figure 17.28 Ovary.
1. Primordial follicles   3. Atretic follicle
2. Primary follicles   4. Tunical albuginea

Figure 17.29 Secondary follicle.
1. Granulosa cells   4. Oocyte
2. Antrum   5. Theca interna
3. Zona pellucida   6. Theca externa

Figure 17.30 Corpus luteum.
1. Corpus luteum   3. Former follicular antrum
2. Wall of corpus luteum

Levels of luteinizing hormone (LH) and follicle stimulating hormone (FSH) in blood plasma

LH

FSH

Levels of estrogen and progesterone in blood plasma

Estrogen

Progesterone

Menstruation cycle as shown by thickness of the endometrium

Ovarian cycle

Follicular cycle

Ovulation

Luteal cycle

Days   1   7   14   21   28

Figure 17.31 The female ovarian and menstruation cycle.

163

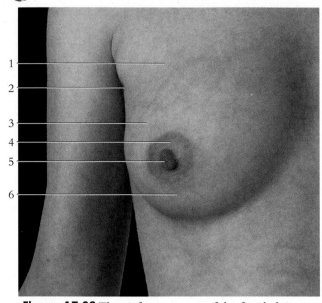

Figure 17.32 The surface anatomy of the female breast.
1. Pectoralis major muscle
2. Axilla
3. Lateral process of breast
4. Areola
5. Nipple
6. Breast (containing mammary glands)

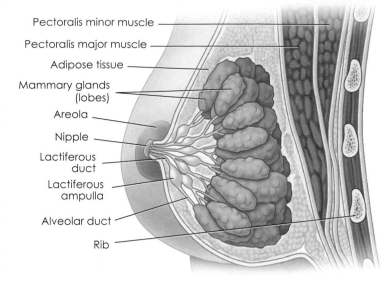

Pectoralis minor muscle
Pectoralis major muscle
Adipose tissue
Mammary glands (lobes)
Areola
Nipple
Lactiferous duct
Lactiferous ampulla
Alveolar duct
Rib

Figure 17.33 The mammary gland.

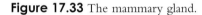

Figure 17.34 Mammary glands, (non-lactating glands).
1. Interlobular duct
2. Interlobular connective tissue
3. Lobule of glandular tissue

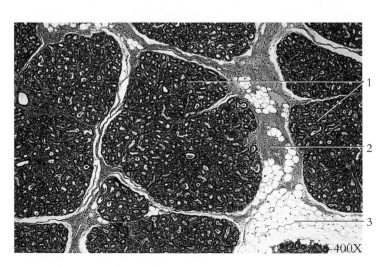

Figure 17.35 Mammary glands, (lactating glands).
1. Lobules of glandular tissue
2. Intralobular connective tissue
3. Adipose cells

Developmental Biology

Chapter 18

The period of human pregnancy, which generally requires 38 weeks, is known as **gestation**. **Morphogenesis** is the sequence of changes that occur in the formation of the baby's body structures. Although gestation is frequently discussed chronologically as trimesters, prenatal development is more accurately divided morphogenically into three periods based on structural changes. On this basis, the **pre-embryonic period** includes the first two weeks following fertilization, the **embryonic period** includes the following six weeks, and the **fetal period** includes the final 30 weeks.

The events of the two-week pre-embryonic period include transportation of the fertilized egg, or **zygote**, through the uterine tube, mitotic divisions, implantation, and the formation of primordial embryonic tissue (Fig. 18.1). Implantation begins between the fifth and seventh day and is made possible by the secretion of enzymes that digest a portion of the endometrium of the uterus. During implantation, the **trophoblast cells** secrete human chorionic gonadotropin (hCG), which prevents the breakdown of the endometrium and menstruation by maintaining the corpus luteum. The trophoblast cells also participate in the formation of the placenta.

The events of the six-week embryonic period include the differentiation of the germ layers into specific body organs and the formation of the extraembryonic membranes, including the **placenta, umbilical cord, amnion, yolk sac, allantois**, and **chorion**.

A small amount of tissue differentiation and organ development occurs during the fetal period, but for the most part fetal development is limited to body growth. Labor and parturition (childbirth) are the culmination of gestation and require the action of **oxytocin** provided by the hypothalamus and released from the posterior pituitary gland, and **prostaglandins**, produced in her uterus.

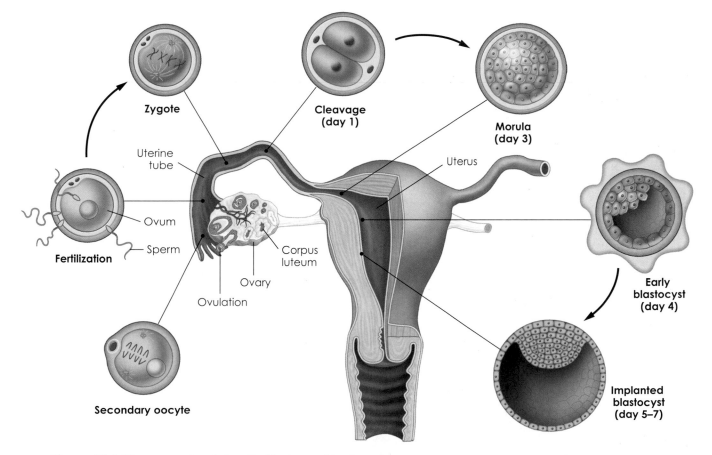

Figure 18.1 The events of ovulation, fertilization, and implantation.

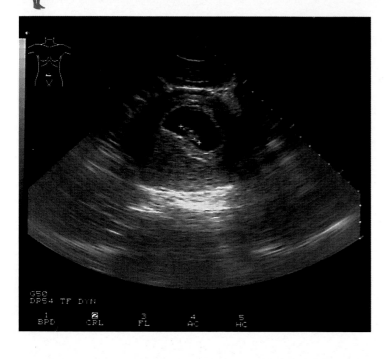

Figure 18.2 A 8.1 week intrauterine pregnancy. Marks are indicated on the image to denote the crown and rump. The crown-rump length of this embryo is 18 mm. Ultrasonography, produced by a mechanical vibration of high frequency, produces a safe, high resolution of fetal structure. Most ultrasound scans are obtained on fetuses older than 12 weeks.

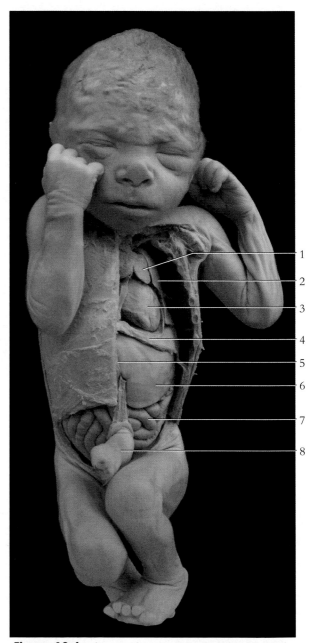

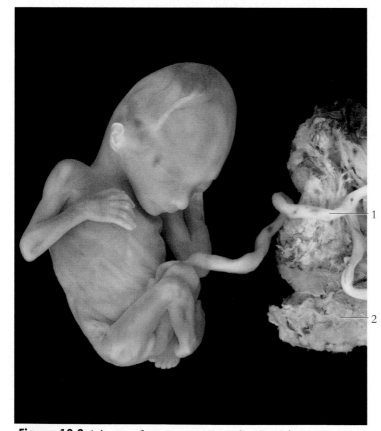

Figure 18.3 A human fetus at approximately 18 weeks.
1. Umbilical cord
2. Placenta

Figure 18.4 A human fetus at approximately 28 weeks.

1. Thymus	5. Falciform ligament
2. Lung	6. Liver
3. Heart	7. Small intestine
4. Diaphragm	8. Umbilical cord

Figure 18.5 The thoracic and abdominal viscera of a human fetus at approximately 28 weeks.

1. Pectoralis major m.
2. Pericardium (cut)
3. Pericardial cavity
4. Falciform ligament
5. External abdominal oblique m.
6. Umbilicus
7. Umbilical vein
8. Thymus
9. Lung
10. Pleural cavity
11. Heart
12. Thoracic wall
13. Diaphragm
14. Liver
15. Peritoneal cavity
16. Abdominal wall
17. Small intestine
18. Umbilical cord

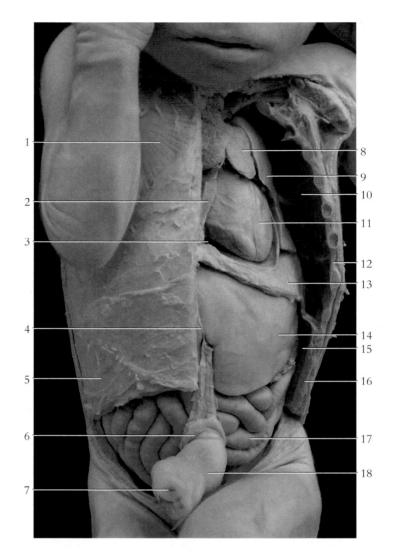

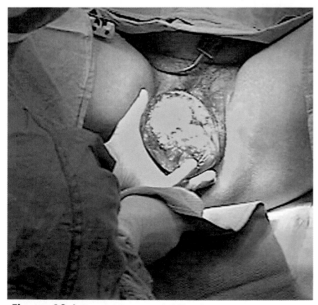

Figure 18.6 Parturition, or childbirth.

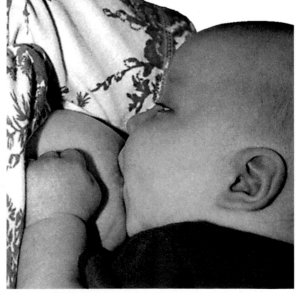

Figure 18.7 Nursing.

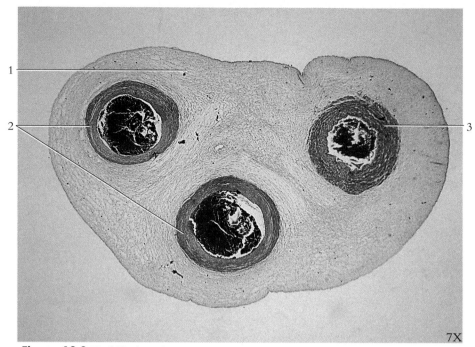

Figure 18.8 Umbilical cord.
1. Mucus connective tissue
   (Wharton's jelly)
2. Umbilical arteries
3. Umbilical vein

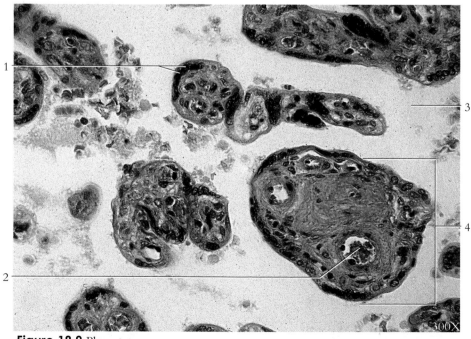

Figure 18.9 Placenta.
1. Nuclei of syncytiotrophoblast
2. Blood vessel containing fetal
   blood cells
3. Intervillus space containing
   maternal blood cells
4. Chorionic villus

Cat Dissection

Chapter 19

Embalmed cats purchased from biological supply houses are excellent specimens for dissecting and learning basic mammalian anatomy. Before the muscles and viscera of a cat can be studied, the specimen's skin has to be removed according to the following suggested guidelines.

1. Place the cat on a dissecting tray dorsal side up. Using a sharp scalpel, make a short, shallow incision through the skin across the nape of the neck. With your scissors, continue a dorsal midline incision forward over the skull and down the back to about two inches onto the tail. Sever the tail with bone shears or a saw and discard.
2. Make a shallow incision around the neck and down each foreleg to the paws. Continue a circular incision around each wrist. Beginning at the base of the tail, make incisions down each of the hind legs to the ankles. Make a circular cut around each ankle.
3. Carefully remove the skin, using your fingers or a blunt probe to separate the skin from underlying connective tissue. Where it is necessary to use a scalpel, keep the cutting edge directed toward the skin away from the muscle. If your specimen is a male, make an incision around the genitalia, leaving the skin intact. If your specimen is a female, the mammary glands will appear as longitudinal, glandular masses along the ventral side of the abdomen and thorax. They should be removed with the skin.

4. After the specimen is skinned, remove the excess fat and connective tissue to expose the underlying muscles. Make certain that the muscles are separated along their natural boundaries. If a transection of a muscle is necessary, isolate the muscle from its attached connective tissue and make a clean cut across the belly of the muscle, leaving the origin and insertion intact.
5. At the end of the laboratory period, wrap your specimen in muslin cloth and store it in a tight, heavy-duty plastic bag. Wet your specimen from time to time with a preservative solution (usually 2%-3% phenol). Caution needs to be taken when using a phenol wetting solution as it is caustic and poisonous if misused or used in a concentrated form.

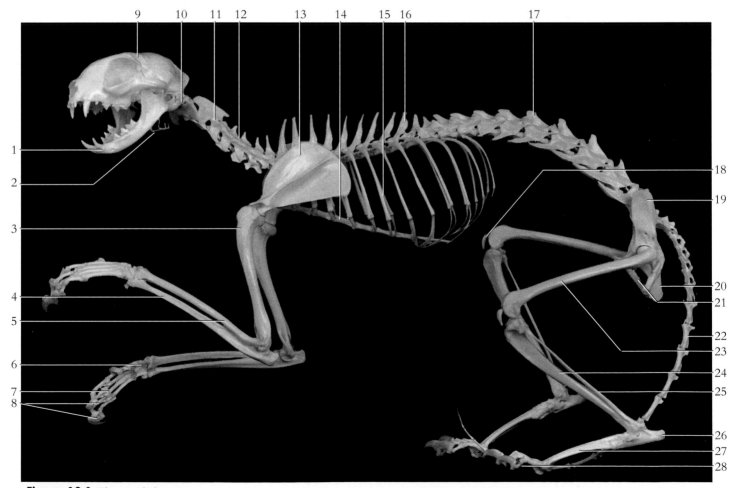

Figure 19.1 The cat skeleton.

1. Mandible
2. Hyoid bone
3. Humerus
4. Ulna
5. Radius
6. Carpal bones
7. Metacarpal bones
8. Phalanges
9. Skull
10. Atlas
11. Axis
12. Cervical vertebra
13. Scapula
14. Sternum
15. Rib
16. Thoracic vertebra
17. Lumbar vertebra
18. Patella
19. Ilium
20. Ischium
21. Pubis
22. Caudal vertebra
23. Femur
24. Tibia
25. Fibula
26. Tarsal bones
27. Metatarsal bones
28. Phalanges

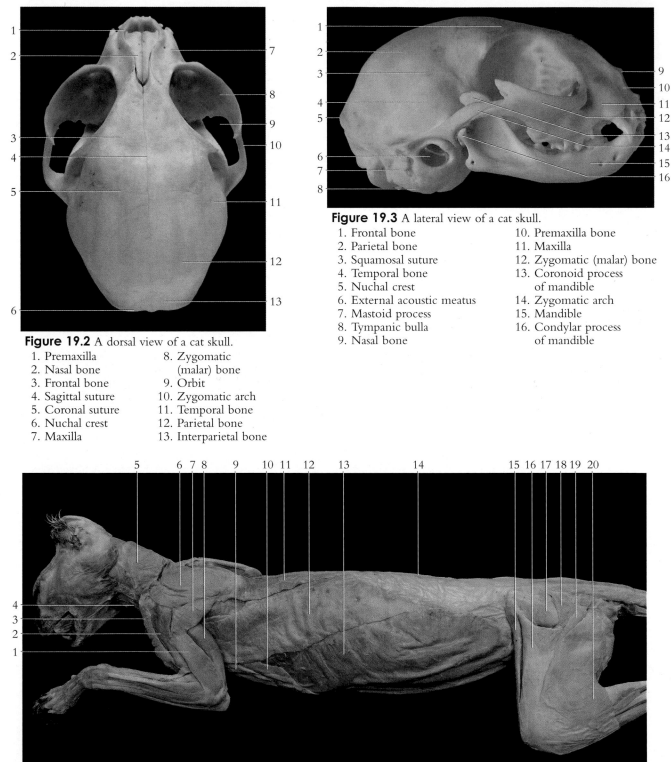

Figure 19.2 A dorsal view of a cat skull.

1. Premaxilla
2. Nasal bone
3. Frontal bone
4. Sagittal suture
5. Coronal suture
6. Nuchal crest
7. Maxilla
8. Zygomatic (malar) bone
9. Orbit
10. Zygomatic arch
11. Temporal bone
12. Parietal bone
13. Interparietal bone

Figure 19.3 A lateral view of a cat skull.

1. Frontal bone
2. Parietal bone
3. Squamosal suture
4. Temporal bone
5. Nuchal crest
6. External acoustic meatus
7. Mastoid process
8. Tympanic bulla
9. Nasal bone
10. Premaxilla bone
11. Maxilla
12. Zygomatic (malar) bone
13. Coronoid process of mandible
14. Zygomatic arch
15. Mandible
16. Condylar process of mandible

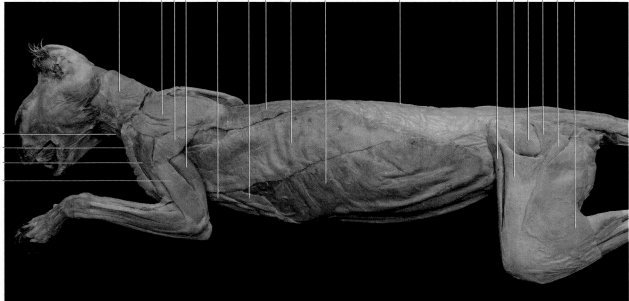

Figure 19.4 A lateral view of the superficial muscles of the cat.

1. Lateral head of triceps brachii m.
2. Acromiodeltoid m.
3. Clavobrachialis m. (clavodeltiod)
4. Sternomastoid m.
5. Clavotrapezius m.
6. Acromiotrapezius m.
7. Spinodeltoid m.
8. Long head of triceps brachii m.
9. Pectoralis minor m.
10. Xiphihumeralis m.
11. Spinotrapezius m.
12. Latissimus dorsi m.
13. External abdominal oblique m.
14. Lumbodorsal fascia
15. Sartorius m.
16. Tensor fasciae latae m.
17. Gluteus medius m.
18. Gluteus maximus m.
19. Caudofemoralis m.
20. Biceps femoris m.

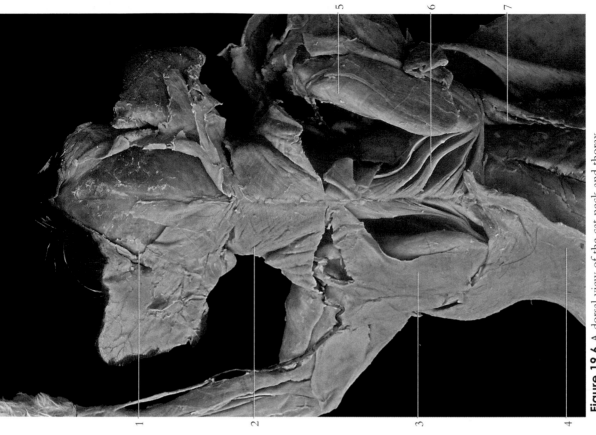

Figure 19.6 A dorsal view of the cat neck and thorax.

1. Temporalis m.
2. Clavotrapezius m.
3. Acromiotrapezius m.
4. Latissimus dorsi m.
5. Supraspinatus m.
6. Rhomboid m.
7. Serratus anterior m.

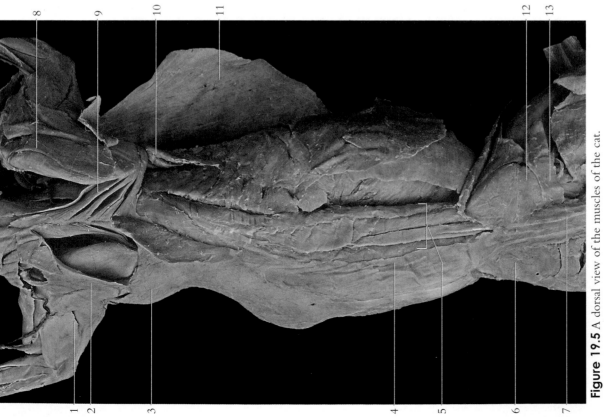

Figure 19.5 A dorsal view of the muscles of the cat.

1. Lateral head of triceps brachii m.
2. Acromiotrapezius m.
3. Latissimus dorsi m.
4. Lumbodorsal fascia
5. Sacrospinalis m.
6. Gluteus medius m.
7. Caudal m.
8. Supraspinatus m.
9. Rhomboid m.
10. Serratus anterior m.
11. Latissimus dorsi m.
12. Gluteus maximus m.
13. Caudofemoralis m.

171

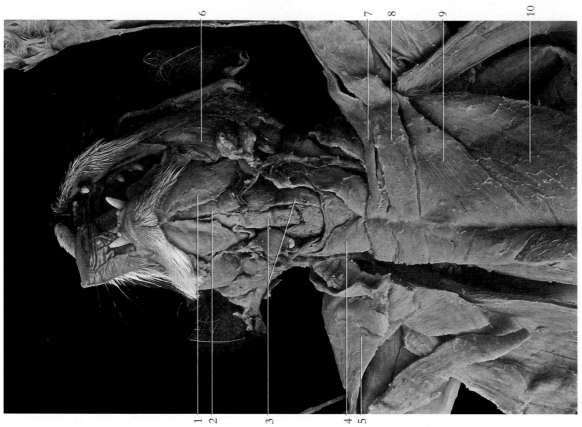

Figure 19.8 A ventral view of the neck and thorax.

1. Digastric m.
2. Mylohyoid m.
3. Sternohyoid m.
4. Sternomastoid m.
5. Clavotrapezius m.
6. Masseter m.
7. Clavobrachialis m. (clavodeltiod)
8. Pectoantebrachialis m.
9. Pectoralis major m.
10. Pectoralis minor m.

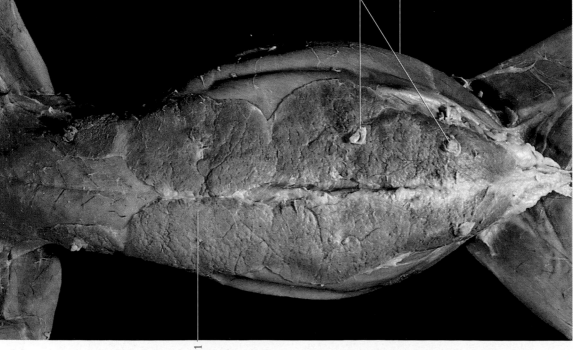

Figure 19.7 A superficial view of the ventral trunk.

1. Mammary glands
2. Nipples
3. External abdominal oblique m.

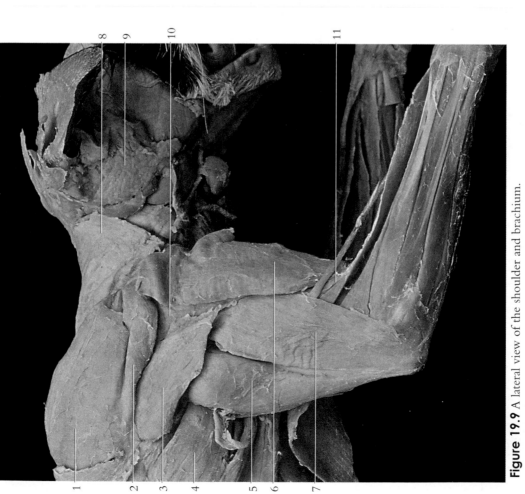

Figure 19.10 An anterior view of the brachium and antebrachium.

1. Extensor carpi radialis longus m.
2. Brachioradialis m.
3. Palmaris longus m. (cut)
4. Flexor carpi ulnaris m.
5. Pronator teres m.
6. Epitrochlearis
7. Masseter m.
8. Sternomastoid m.
9. Clavobrachialis m. (clavodeltiod)
10. Pectoantebrachialis m.
11. Pectoralis major m.
12. Pectoralis minor m.

Figure 19.9 A lateral view of the shoulder and brachium.

1. Acromiotrapezius m.
2. Levator scapulae ventralis m.
3. Spinodeltoid m.
4. Latissimus dorsi m.
5. Long head of triceps brachii m.
6. Clavobrachialis m. (clavodeltiod)
7. Lateral head of triceps brachii m.
8. Clavotrapezius m.
9. Parotid gland
10. Acromiodeltoid m.
11. Brachioradialis m.

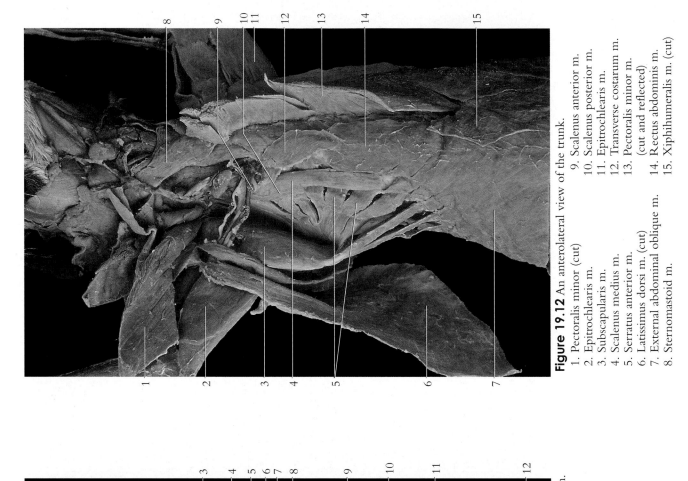

Figure 19.12 An anterolateral view of the trunk.

1. Pectoralis minor (cut)
2. Epitrochlearis m.
3. Subscapularis m.
4. Scalenus medius m.
5. Serratus anterior m.
6. Latissimus dorsi m. (cut)
7. External abdominal oblique m.
8. Sternomastoid m.
9. Scalenus anterior m.
10. Scalenus posterior m.
11. Epitrochlearis m.
12. Transverse costarum m.
13. Pectoralis minor m. (cut and reflected)
14. Rectus abdominis m.
15. Xiphihumeralis m. (cut)

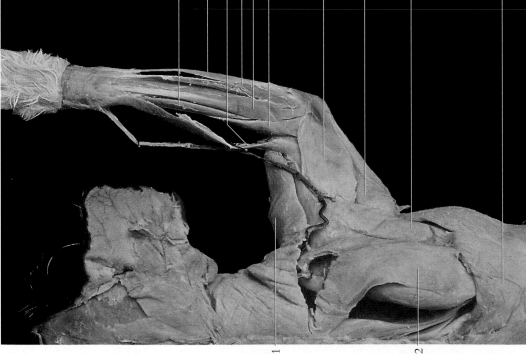

Figure 19.11 A posterior view of the brachium and antebrachium.

1. Clavobrachialis m.
2. Acromiotrapezius m.
3. Extensor carpi radialis brevis m.
4. Extensor carpi ulnaris m.
5. Brachioradialis m.
6. Extensor digitorum lateralis m.
7. Extensor digitorum communis m.
8. Extensor carpi radialis longus m.
9. Lateral head of triceps brachii m.
10. Long head of triceps brachii m.
11. Spinodeltoid m.
12. Latissimus dorsi m.

Figure 19.13 A lateral view of the trunk.

1. Internal abdominal oblique m.
2. Tensor fascia latae
3. Caudofemoris m.
4. Vastus lateralis m.
5. Sartorius m.
6. External abdominal oblique m. (cut and reflected)
7. Latissimus dorsi m.
8. Spinodeltoid m.
9. Transverse abdominis m.
10. Serratus anterior m.
11. Long head of triceps brachii m.

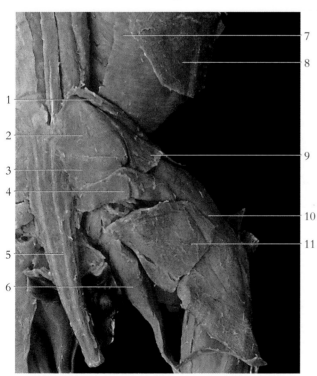

Figure 19.14 A lateral view of the superficial thigh.

1. Sartorius m.
2. Gluteus medius m.
3. Gluteus maximus m.
4. Caudofemoralis m.
5. Caudal m.
6. Semitendinosus m.
7. Internal abdominal oblique m.
8. External abdominal oblique m.
9. Tensor fascia latae (cut)
10. Vastus lateralis m.
11. Biceps femoris m.

Figure 19.15 A medial view of the thigh and leg.

1. Sartorius m. (cut)
2. Vastus lateralis m.
3. Rectus femoris m.
4. Vastus medialis m.
5. Flexor digitorum longus m.
6. Tibialis anterior m.
7. Rectus abdominus m.
8. Adductor longus m.
9. Adductor femoris m.
10. Semimembranosus m.
11. Gracilis m. (cut)
12. Gastrocnemius m.
13. Tendo calcaneus (Achilles tendon)

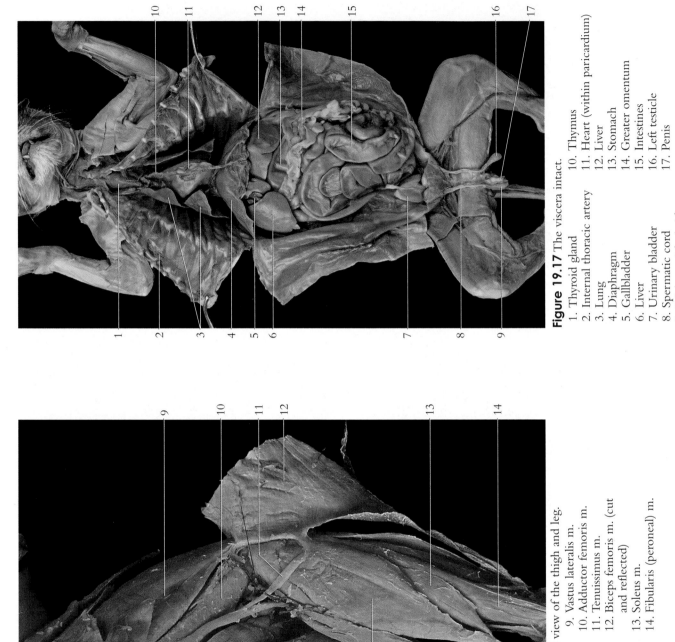

Figure 19.17 The viscera intact.

1. Thyroid gland
2. Internal thoracic artery
3. Lung
4. Diaphragm
5. Gallbladder
6. Liver
7. Urinary bladder
8. Spermatic cord
9. Right testicle (with partial scrotum)
10. Thymus
11. Heart (within paricardium)
12. Liver
13. Stomach
14. Greater omentum
15. Intestines
16. Left testicle
17. Penis

Figure 19.16 A posterolateral view of the thigh and leg.

1. Gluteus medius m.
2. Gluteus maximus m.
3. Caudofemoralis m.
4. Sciatic nerve
5. Semimembranosus m.
6. Semitendinosus m.
7. Gastrocnemius m.
8. Tendo calcaneus
9. Vastus lateralis m.
10. Adductor femoris m.
11. Tenuissimus m.
12. Biceps femoris m. (cut and reflected)
13. Soleus m.
14. Fibularis (peroneal) m.

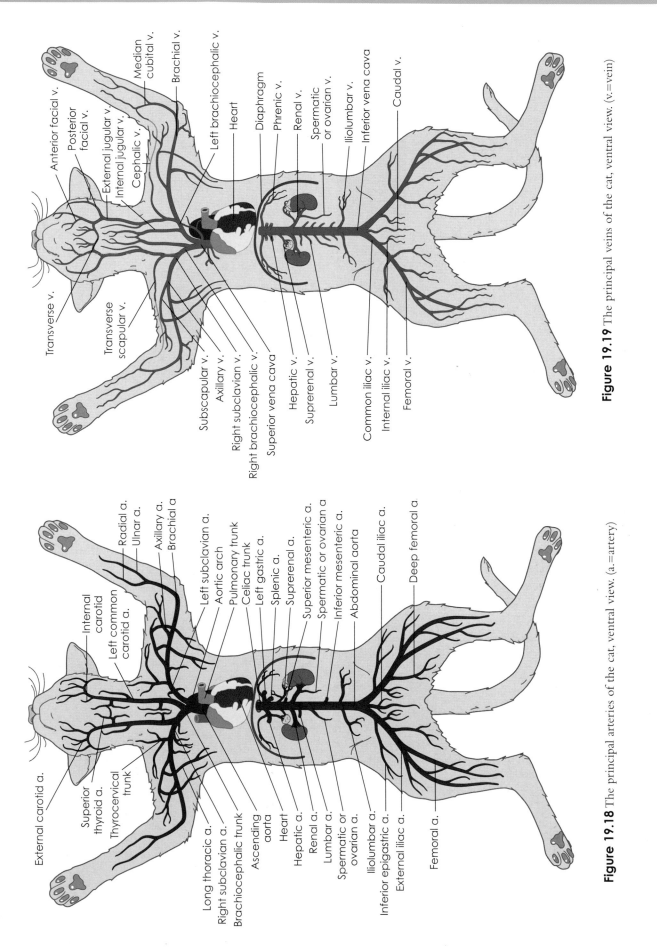

Figure 19.19 The principal veins of the cat, ventral view. (v.=vein)

Figure 19.18 The principal arteries of the cat, ventral view. (a.=artery)

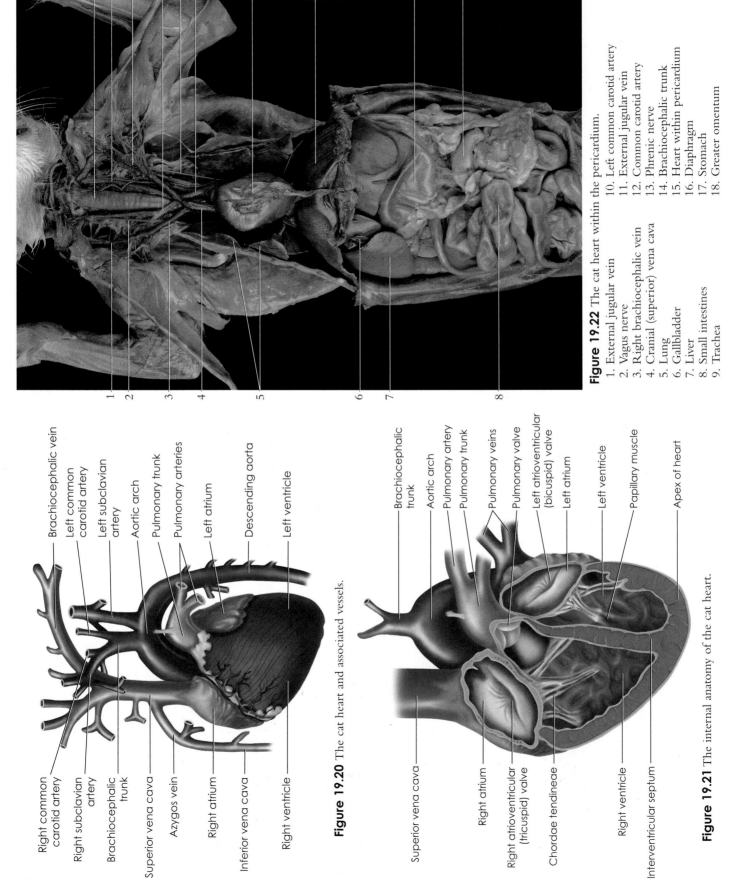

Figure 19.22 The cat heart within the pericardium.

1. External jugular vein
2. Vagus nerve
3. Right brachiocephalic vein
4. Cranial (superior) vena cava
5. Lung
6. Gallbladder
7. Liver
8. Small intestines
9. Trachea
10. Left common carotid artery
11. External jugular vein
12. Common carotid artery
13. Phrenic nerve
14. Brachiocephalic trunk
15. Heart within pericardium
16. Diaphragm
17. Stomach
18. Greater omentum

Right common carotid artery
Right subclavian artery
Brachiocephalic trunk
Superior vena cava
Azygos vein
Right atrium
Inferior vena cava
Right ventricle

Brachiocephalic vein
Left common carotid artery
Left subclavian artery
Aortic arch
Pulmonary trunk
Pulmonary arteries
Left atrium
Descending aorta
Left ventricle

Figure 19.20 The cat heart and associated vessels.

Superior vena cava
Right atrium
Right atrioventricular (tricuspid) valve
Chordae tendineae
Right ventricle
Interventricular septum

Brachiocephalic trunk
Aortic arch
Pulmonary artery
Pulmonary trunk
Pulmonary veins
Pulmonary valve
Left atrioventricular (bicuspid) valve
Left atrium
Left ventricle
Papillary muscle
Apex of heart

Figure 19.21 The internal anatomy of the cat heart.

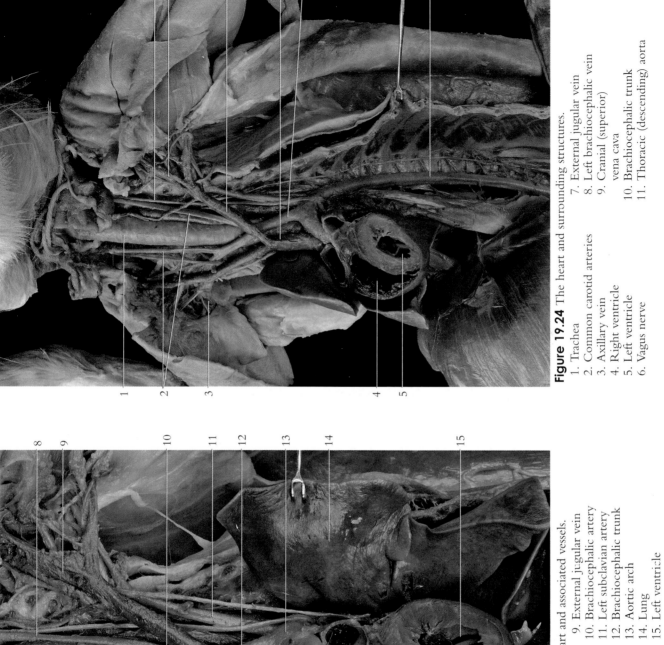

Figure 19.24 The heart and surrounding structures.
1. Trachea
2. Common carotid arteries
3. Axillary vein
4. Right ventricle
5. Left ventricle
6. Vagus nerve
7. External jugular vein
8. Left brachiocephalic vein
9. Cranial (superior) vena cava
10. Brachiocephalic trunk
11. Thoracic (descending) aorta

Figure 19.23 An internal view of the heart and associated vessels.
1. Trachea
2. Right common carotid artery
3. Right subclavian vein
4. Right brachiocephalic vein
5. Cranial (superior) vena cava
6. Right atrium
7. Right ventricle
8. Left common carotid artery
9. External jugular vein
10. Brachiocephalic artery
11. Left subclavian artery
12. Brachiocephalic trunk
13. Aortic arch
14. Lung
15. Left ventricle

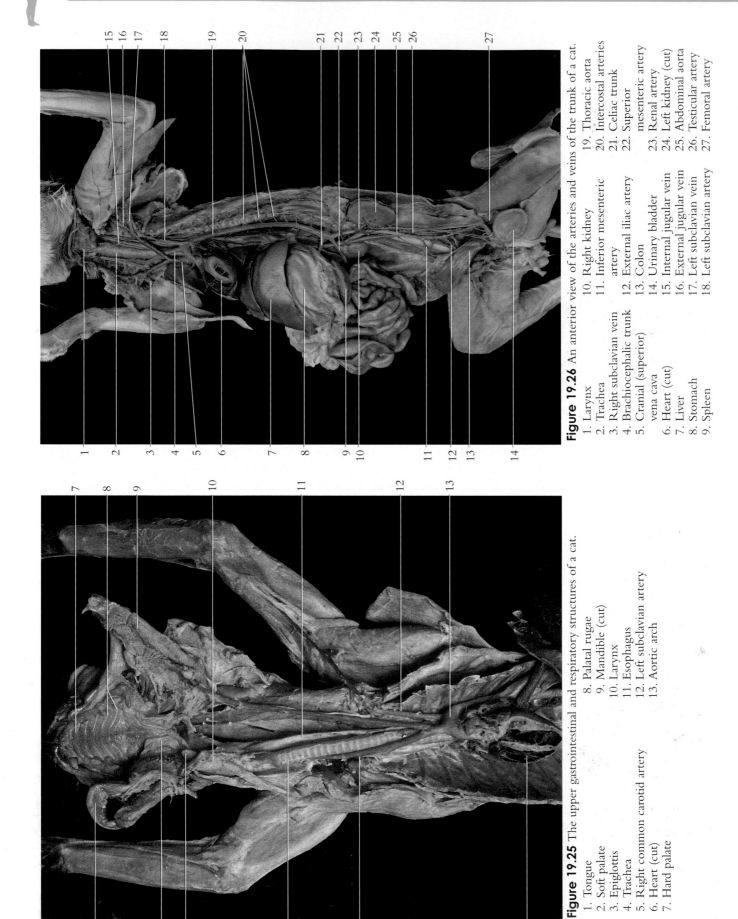

Figure 19.25 The upper gastrointestinal and respiratory structures of a cat.

1. Tongue
2. Soft palate
3. Epiglottis
4. Trachea
5. Right common carotid artery
6. Heart (cut)
7. Hard palate
8. Palatal rugae
9. Mandible (cut)
10. Larynx
11. Esophagus
12. Left subclavian artery
13. Aortic arch

Figure 19.26 An anterior view of the arteries and veins of the trunk of a cat.

1. Larynx
2. Trachea
3. Right subclavian vein
4. Brachiocephalic trunk
5. Cranial (superior) vena cava
6. Heart (cut)
7. Liver
8. Stomach
9. Spleen
10. Right kidney
11. Inferior mesenteric artery
12. External iliac artery
13. Colon
14. Urinary bladder
15. Internal jugular vein
16. External jugular vein
17. Left subclavian vein
18. Left subclavian artery
19. Thoracic aorta
20. Intercostal arteries
21. Celiac trunk
22. Superior mesenteric artery
23. Renal artery
24. Left kidney (cut)
25. Abdominal aorta
26. Testicular artery
27. Femoral artery

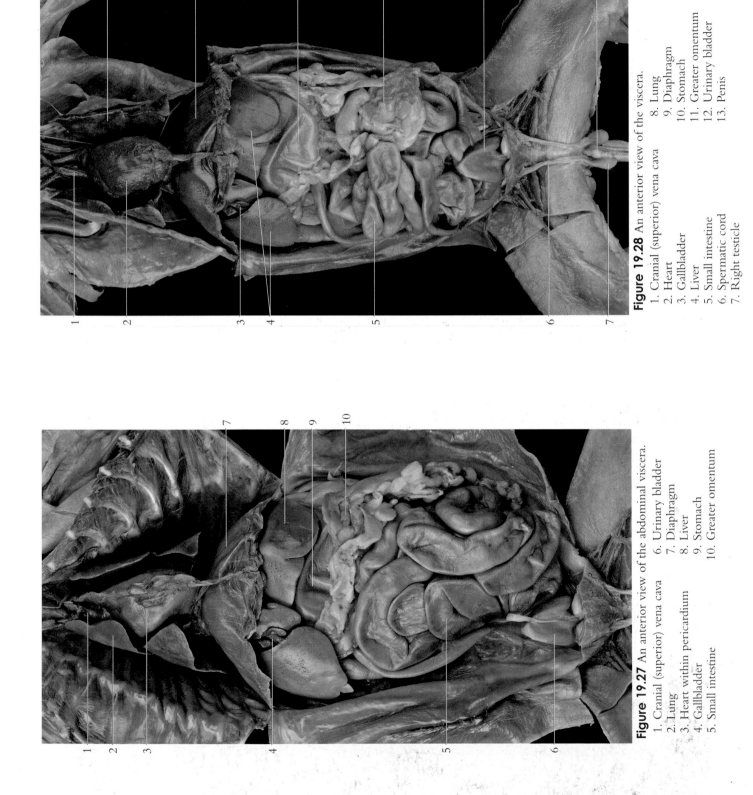

Figure 19.28 An anterior view of the viscera.
1. Cranial (superior) vena cava
2. Heart
3. Gallbladder
4. Liver
5. Small intestine
6. Spermatic cord
7. Right testicle
8. Lung
9. Diaphragm
10. Stomach
11. Greater omentum
12. Urinary bladder
13. Penis

Figure 19.27 An anterior view of the abdominal viscera.
1. Cranial (superior) vena cava
2. Lung
3. Heart within pericardium
4. Gallbladder
5. Small intestine
6. Urinary bladder
7. Diaphragm
8. Liver
9. Stomach
10. Greater omentum

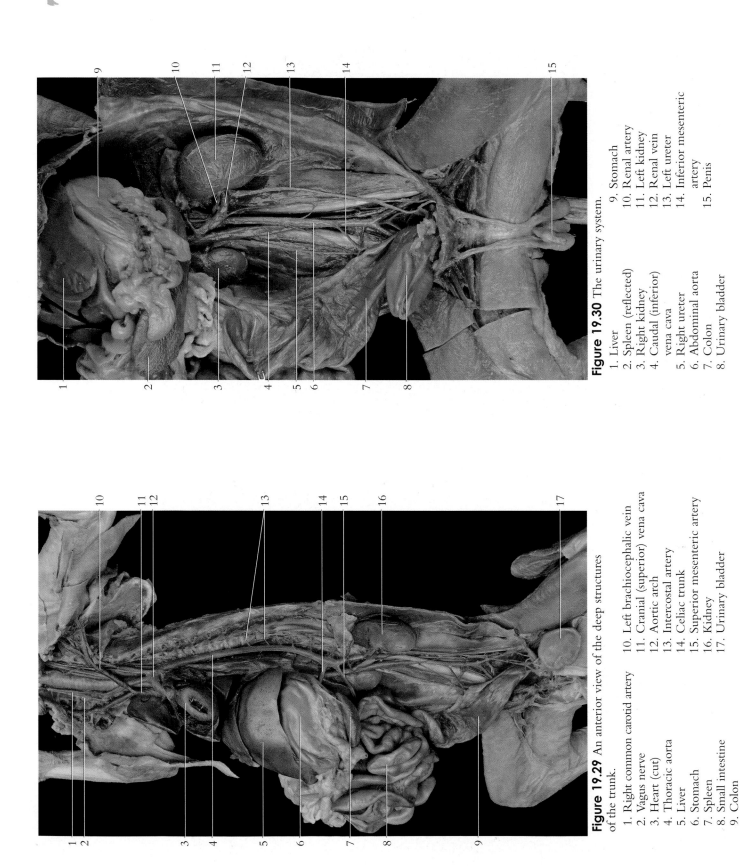

Figure 19.30 The urinary system.

1. Liver
2. Spleen (reflected)
3. Right kidney
4. Caudal (inferior) vena cava
5. Right ureter
6. Abdominal aorta
7. Colon
8. Urinary bladder
9. Stomach
10. Renal artery
11. Left kidney
12. Renal vein
13. Left ureter
14. Inferior mesenteric artery
15. Penis

Figure 19.29 An anterior view of the deep structures of the trunk.

1. Right common carotid artery
2. Vagus nerve
3. Heart (cut)
4. Thoracic aorta
5. Liver
6. Stomach
7. Spleen
8. Small intestine
9. Colon
10. Left brachiocephalic vein
11. Cranial (superior) vena cava
12. Aortic arch
13. Intercostal artery
14. Celiac trunk
15. Superior mesenteric artery
16. Kidney
17. Urinary bladder

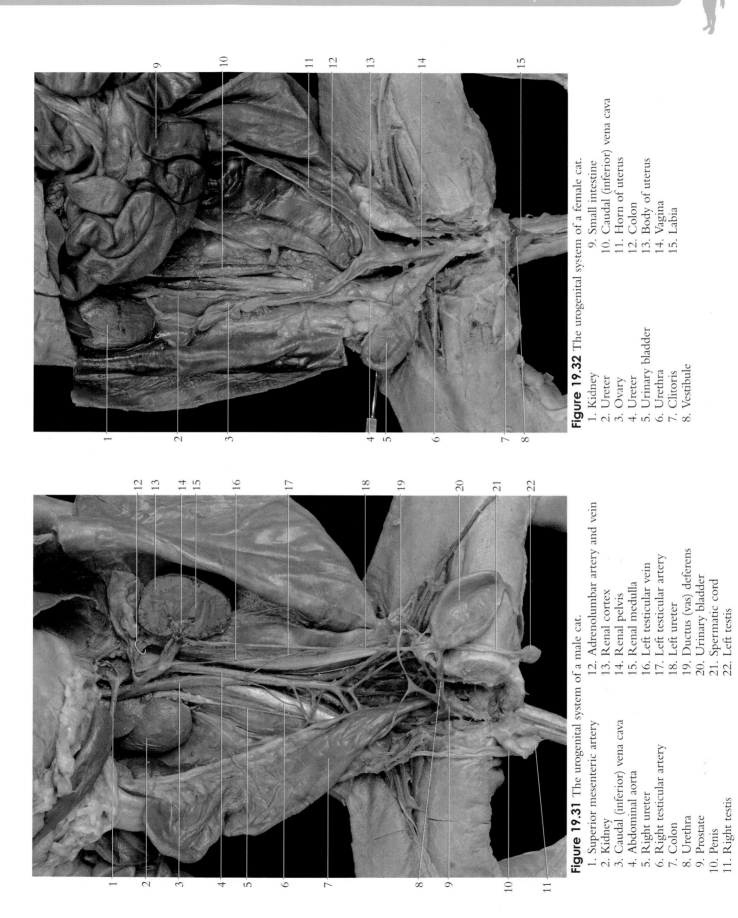

Figure 19.32 The urogenital system of a female cat.

1. Kidney
2. Ureter
3. Ovary
4. Ureter
5. Urinary bladder
6. Urethra
7. Clitoris
8. Vestibule
9. Small intestine
10. Caudal (inferior) vena cava
11. Horn of uterus
12. Colon
13. Body of uterus
14. Vagina
15. Labia

Figure 19.31 The urogenital system of a male cat.

1. Superior mesenteric artery
2. Kidney
3. Caudal (inferior) vena cava
4. Abdominal aorta
5. Right ureter
6. Right testicular artery
7. Colon
8. Urethra
9. Prostate
10. Penis
11. Right testis
12. Adrenolumbar artery and vein
13. Renal cortex
14. Renal pelvis
15. Renal medulla
16. Left testicular vein
17. Left testicular artery
18. Left ureter
19. Ductus (vas) deferens
20. Urinary bladder
21. Spermatic cord
22. Left testis

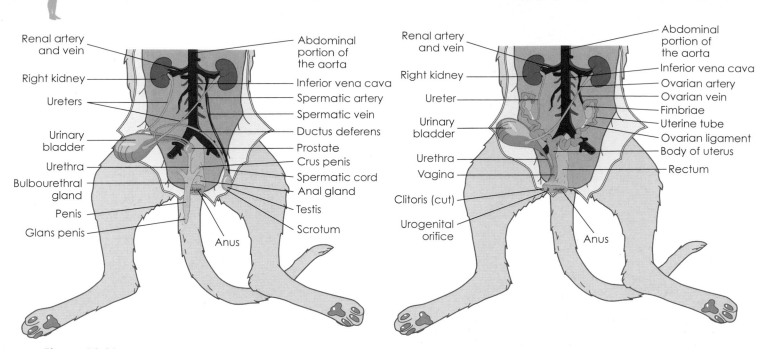

Figure 19.33 A diagram of the urogenital system of a male cat.

Figure 19.34 A diagram of the urogenital system of a female cat.

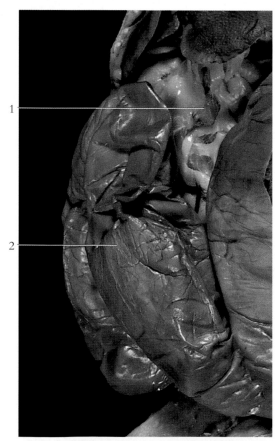

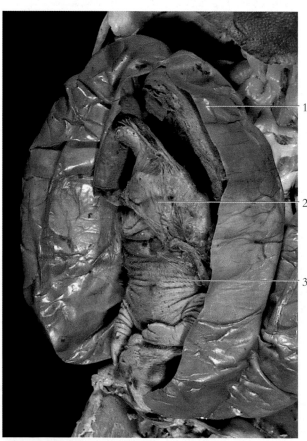

Figure 19.35 The abdominal cavity of a pregnant cat.
1. Greater omentum
2. Right horn of uterus

Figure 19.36 The uterine horn of a pregnant cat.
1. Uterine wall (cut)
2. Amniotic sac enclosing fetus
3. Fetus

Fetal Pig Dissection

Because much can be learned from dissecting embalmed fetal pig specimens, they are frequently utilized in anatomy laboratories. Fetal pigs are purchased from biological supply houses and are specially prepared for dissection. Excess embalming fluid should be drained from the packaged specimen prior to dissection.

Examine your specimen and identify the **umbilical cord** attached to the ventral surface of the abdomen. Locate the two rows of **teats** that extend the length of the abdomen. Determine the sex of your specimen. A male has a **scrotal sac** in the pelvic region of the body between the hind legs and a **urogenital opening** just caudal to the umbilical cord. The **penis** can be palpated as a muscular tubular structure just underneath the skin along the midline proceeding caudally from the urogenital opening. A female has a small fleshy **genital papilla** projecting from the urogenital opening, which is located immediately ventral to the **anal opening**.

Before the muscles and viscera of a fetal pig can be studied, the specimen's skin has to be removed according to the following suggested guidelines.

1. Place your specimen on a dissecting tray ventral side up. Using a sharp scalpel, make a shallow incision through the skin extending from the chin caudally to the umbilical cord. Carefully continue your cut around one side of the umbilical cord. If your specimen is a male, make a diagonal cut from the umbilical cord to the scrotum. If a female, continue a midventral incision from the umbilical cord to the genital papilla. Make an incision around the genitalia and tail.

2. From the midventral incision, extend an incision down the medial surfaces of the forelegs to the hoofs and then do the same for the skin of the hind legs. Make circular incisions around each of the hoofs. Following the ventral borders of the lower jaws, make extended cuts from the chin dorsolaterally to just below the ears.

3. Grasp the cut edge of the skin and carefully remove it from your specimen. If the skin is difficult to remove, grasp the cut edge of the skin with one hand and push on the muscle with the thumb of the other hand.

4. After the specimen is skinned, the muscles can be seen more easily if the moisture on them is sponged away with a paper towel. The muscles of a fetal pig are extremely delicate and as you proceed to dissect your specimen, make certain that you separate the muscles along their natural boundaries. When transection of a muscle is necessary, carefully isolate the muscle from its attached connective tissue and make a clean cut across the belly of the muscle, leaving the origin and insertion intact.

5. At the end of the laboratory period, wrap your specimen in muslin cloth and store it in a tight, heavy-duty plastic bag. Discard the skin that was removed from your specimen, and the plastic shipment bag. Wet your specimen from time to time with a preservative solution (usually 2 to 3 percent phenol). Caution is necessary when using a phenol wetting solution as it is caustic and poisonous if misused or used in a concentrated form.

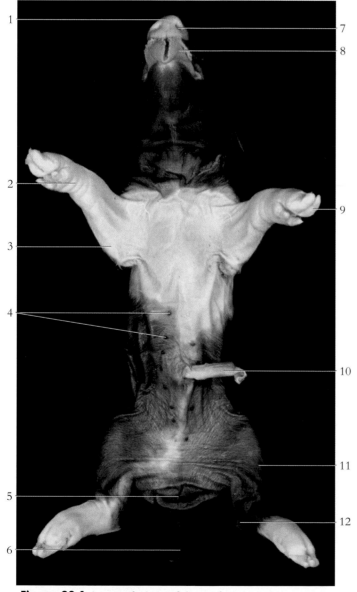

Figure 20.1 A ventral view of the surface anatomy of the male fetal pig.

1. Nose	5. Scrotum	9. Hoof of digit
2. Wrist	6. Tail	10. Umbilical cord
3. Elbow	7. Nostril (nares)	11. Knee
4. Teats	8. Tongue	12. Ankle

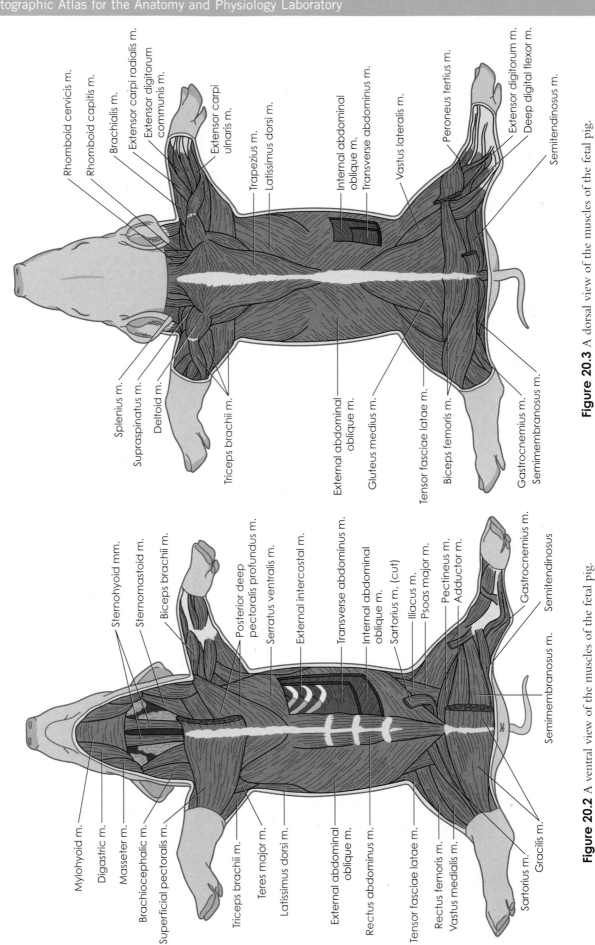

Figure 20.3 A dorsal view of the muscles of the fetal pig.

Figure 20.2 A ventral view of the muscles of the fetal pig.

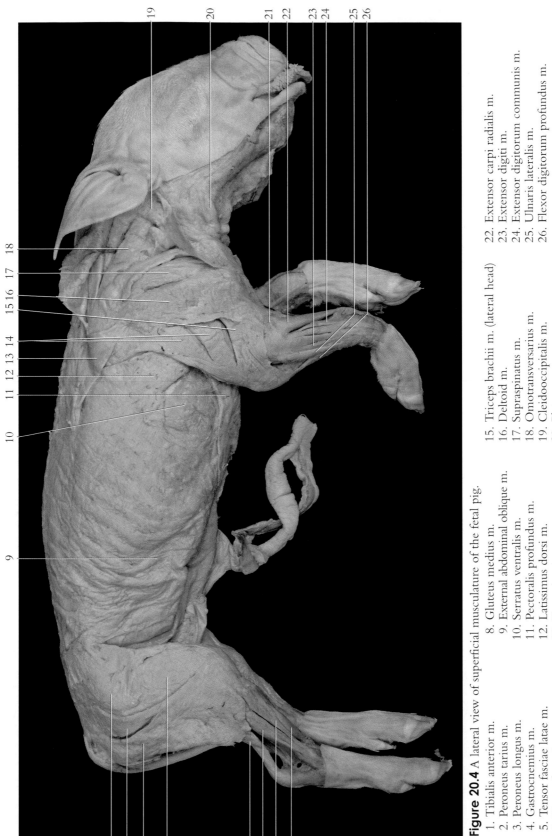

Figure 20.4 A lateral view of superficial musculature of the fetal pig.

1. Tibialis anterior m.
2. Peroneus tarius m.
3. Peroneus longus m.
4. Gastrocnemius m.
5. Tensor fasciae latae m.
6. Biceps femoris m.
7. Gluteus superficialis m.

8. Gluteus medius m.
9. External abdominal oblique m.
10. Serratus ventralis m.
11. Pectoralis profundus m.
12. Latissimus dorsi m.
13. Trapezius m.
14. Triceps brachii m. (long head)

15. Triceps brachii m. (lateral head)
16. Deltoid m.
17. Supraspinatus m.
18. Omotransversarius m.
19. Cleidooccipitalis m.
20. Platysma m.
21. Brachialis m.

22. Extensor carpi radialis m.
23. Extensor digiti m.
24. Extensor digitorum communis m.
25. Ulnaris lateralis m.
26. Flexor digitorum profundus m.

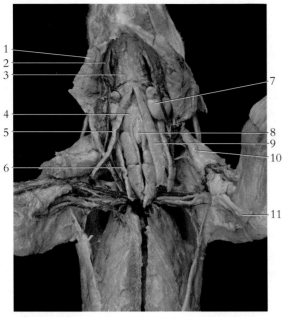

Figure 20.5 A ventral view of superficial muscles of neck and upper torso.

1. Platysma m. (reflected)
2. Digastric m.
3. Mylohyoid m.
4. Sternohyoid m.
5. Omohyoid m.
6. Sternomastoid m.
7. Mandibular gland
8. Larynx
9. Sternothyroid m.
10. Brachiocephalic m.
11. Pectoralis superficialis m. (cut and reflected)

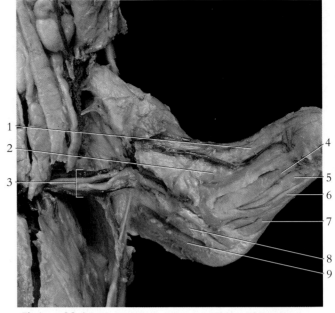

Figure 20.6 Superficial medial muscles of the forelimb.

1. Extensor carpi radialis m.
2. Biceps brachii m.
3. Axillary artery and vein, brachial plexus
4. Flexor carpi radialis m.
5. Flexor digitorum profundus m.
6. Flexor digitorum superficialis m.
7. Flexor carpi ulnaris m.
8. Triceps brachii m. (lateral head)
9. Triceps brachii m. (long head)

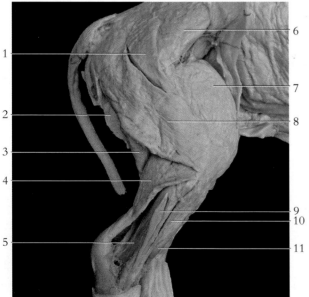

Figure 20.7 A lateral view of the superficial thigh and leg.

1. Gluteus superficialis m.
2. Semitendinosus m.
3. Semimembranosus m.
4. Gastrocnemius m.
5. Extensor digitorum quarti and quinti mm.
6. Gluteus medius m.
7. Tensor fasciae latae m.
8. Biceps femoris m.
9. Fibulais (peroneus) longus m.
10. Fibulais (peroneus) tertius m.
11. Tibialis anterior m.

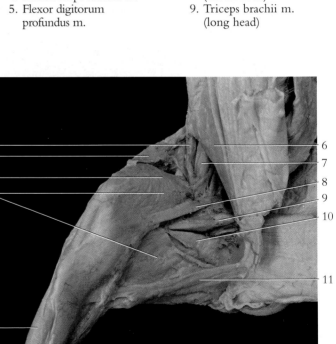
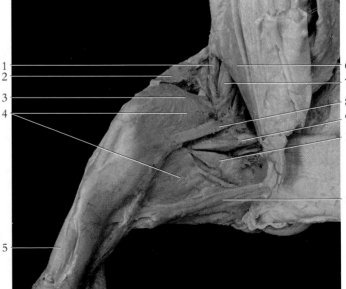

Figure 20.8 Medial muscles of thigh and leg.

1. Iliacus m.
2. Tensor fasciae latae m.
3. Rectus femoris m.
4. Semimembranosus m.
5. Tibialis anterior m.
6. External abdominal oblique m.
7. Psoas major m.
8. Sartorius m.
9. Pectineus m.
10. Adductor m.
11. Semitendinosus m.

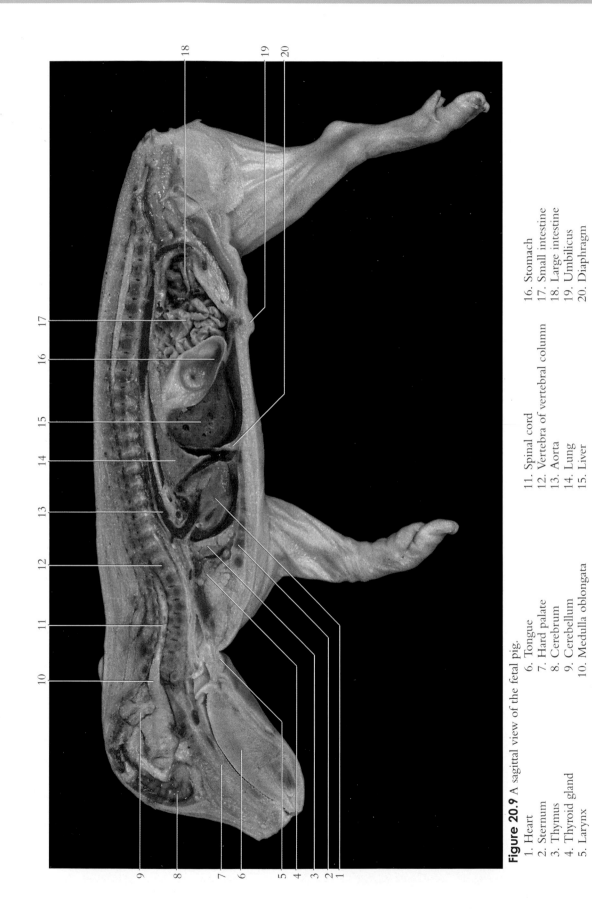

Figure 20.9 A sagittal view of the fetal pig.

1. Heart	6. Tongue	11. Spinal cord	16. Stomach
2. Sternum	7. Hard palate	12. Vertebra of vertebral column	17. Small intestine
3. Thymus	8. Cerebrum	13. Aorta	18. Large intestine
4. Thyroid gland	9. Cerebellum	14. Lung	19. Umbilicus
5. Larynx	10. Medulla oblongata	15. Liver	20. Diaphragm

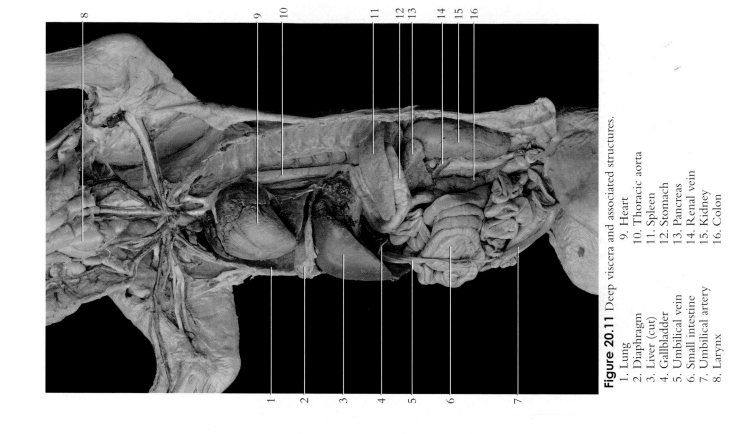

Figure 20.11 Deep viscera and associated structures.

1. Lung
2. Diaphragm
3. Liver (cut)
4. Gallbladder
5. Umbilical vein
6. Small intestine
7. Umbilical artery
8. Larynx
9. Heart
10. Thoracic aorta
11. Spleen
12. Stomach
13. Pancreas
14. Renal vein
15. Kidney
16. Colon

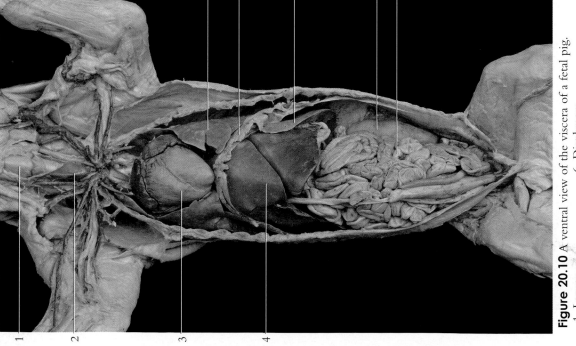

Figure 20.10 A ventral view of the viscera of a fetal pig.

1. Larynx
2. Thyroid gland
3. Heart
4. Liver
5. Lung
6. Diaphragm
7. Spleen
8. Kidney
9. Small intestine

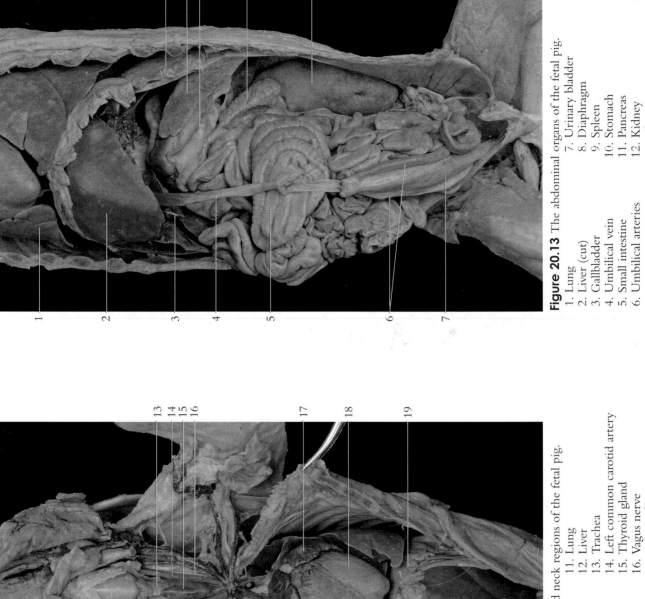

Figure 20.13 The abdominal organs of the fetal pig.

1. Lung
2. Liver (cut)
3. Gallbladder
4. Umbilical vein
5. Small intestine
6. Umbilical arteries
7. Urinary bladder
8. Diaphragm
9. Spleen
10. Stomach
11. Pancreas
12. Kidney

Figure 20.12 The thorax and neck regions of the fetal pig.

1. Larynx
2. Internal jugular vein
3. External jugular vein
4. Subclavian vein
5. Subclavian artery
6. Brachial plexus
7. Internal thoracic artery
8. Internal thoracic vein
9. Right auricle
10. Right ventricle
11. Lung
12. Liver
13. Trachea
14. Left common carotid artery
15. Thyroid gland
16. Vagus nerve
17. Left auricle
18. Left ventricle
19. Diaphragm

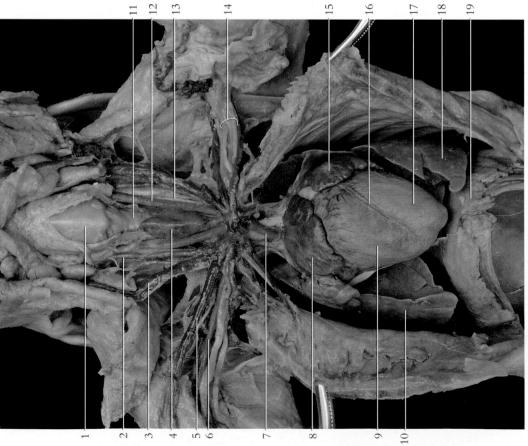

Figure 20.15 Structures of the abdomen and lower extremities.

1. Heart	10. Internal intercostal vessels
2. Lung	11. Spleen
3. Diaphragm	12. Stomach
4. Liver	13. Pancreas
5. Umbilical vein	14. Kidney
6. Small intestine	15. Renal vein
7. Colon	16. Caudal (inferior) vena cava
8. Umbilical artery	17. Renal artery
9. Thoracic aorta	18. Abdominal aorta

Figure 20.14 The arteries and veins of the neck and thoracic region.

1. Larynx	11. Trachea
2. Internal jugular vein	12. Left common carotid artery
3. External jugular vein	13. Vagus nerve
4. Thyroid gland	14. Brachial plexus
5. Subclavian vein	15. Left auricle
6. Subclavian artery	16. Coronary vessels
7. Cranial (superior) vena cava	17. Left ventricle
8. Right auricle	18. Left lung
9. Right ventricle	19. Diaphragm
10. Right lung	

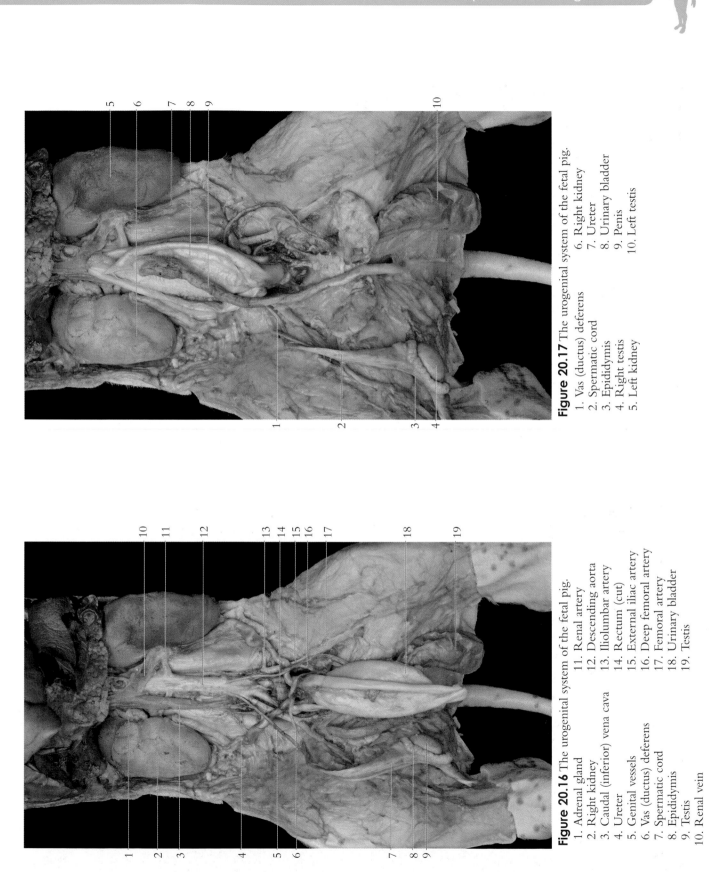

Figure 20.17 The urogenital system of the fetal pig.

1. Vas (ductus) deferens
2. Spermatic cord
3. Epididymis
4. Right testis
5. Left kidney
6. Right kidney
7. Ureter
8. Urinary bladder
9. Penis
10. Left testis

Figure 20.16 The urogenital system of the fetal pig.

1. Adrenal gland
2. Right kidney
3. Caudal (inferior) vena cava
4. Ureter
5. Genital vessels
6. Vas (ductus) deferens
7. Spermatic cord
8. Epididymis
9. Testis
10. Renal vein
11. Renal artery
12. Descending aorta
13. Iliolumbar artery
14. Rectum (cut)
15. External iliac artery
16. Deep femoral artery
17. Femoral artery
18. Urinary bladder
19. Testis

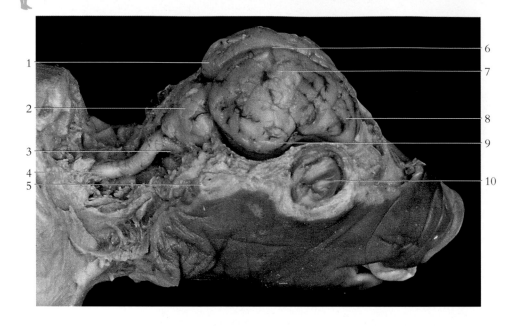

Figure 20.18 The general structures of the fetal pig brain. Because the cerebrum is less defined in pigs, the regions are not known as lobes as they are in humans.

1. Occipital region of cerebrum
2. Cerebellum
3. Medulla oblongata
4. Spinal cord
5. External acoustic meatus
6. Longitudinal fissure
7. Parietal region of cerebrum
8. Frontal region of cerebrum
9. Temporal region of cerebrum
10. Eye

Rat Dissection

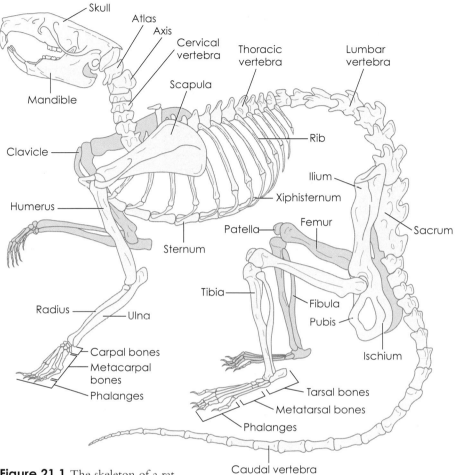

Chapter 21

Figure 21.1 The skeleton of a rat.

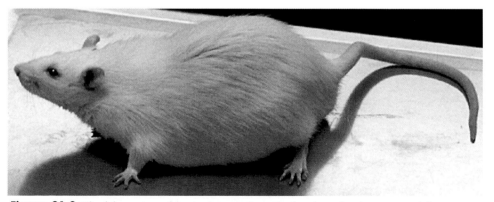

Figure 21.2 The laboratory white rat is a captive-raised rodent that is commercially available for biological and medical experiments and research. White rats are also embalmed and available as dissection specimens in biology, vertebrate biology, and general zoology laboratories.

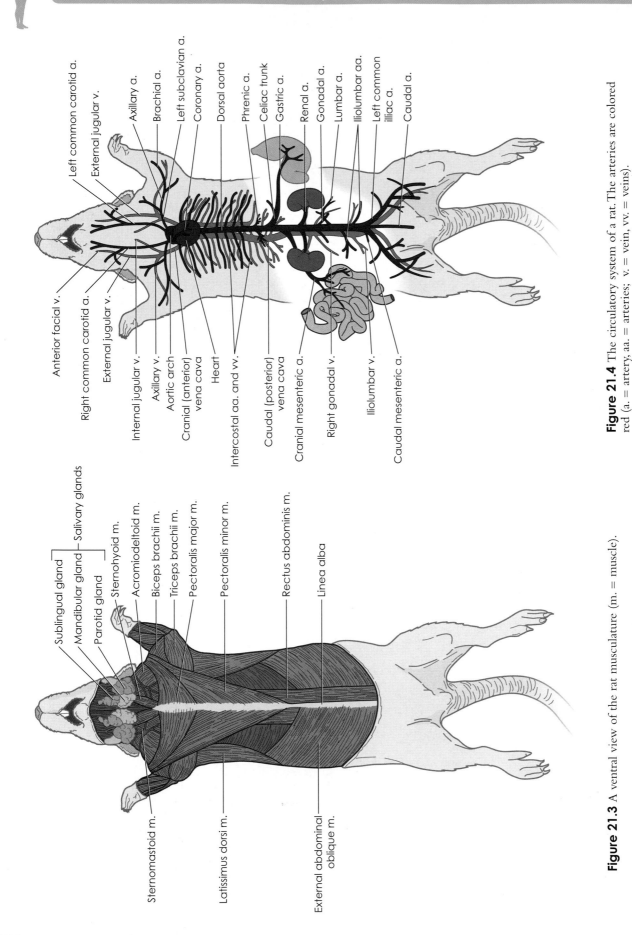

Figure 21.4 The circulatory system of a rat. The arteries are colored red (a. = artery, aa. = arteries; v. = vein, vv. = veins).

Figure 21.3 A ventral view of the rat musculature (m. = muscle).

Left common carotid a.

External jugular v.

Axillary a.

Brachial a.

Left subclavian a.

Coronary a.

Dorsal aorta

Phrenic a.

Celiac trunk

Gastric a.

Renal a.

Gonadal a.

Lumbar a.

Iliolumbar aa.

Left common iliac a.

Caudal a.

Anterior facial v.

Right common carotid a.

External jugular v.

Internal jugular v.

Axillary v.

Aortic arch

Cranial (anterior) vena cava

Heart

Intercostal aa. and vv.

Caudal (posterior) vena cava

Cranial mesenteric a.

Right gonadal v.

Iliolumbar v.

Caudal mesenteric a.

Sublingual gland

Mandibular gland — Salivary glands

Parotid gland

Sternohyoid m.

Acromiodeltoid m.

Biceps brachii m.

Triceps brachii m.

Pectoralis major m.

Pectoralis minor m.

Rectus abdominis m.

Linea alba

Sternomastoid m.

Latissimus dorsi m.

External abdominal oblique m.

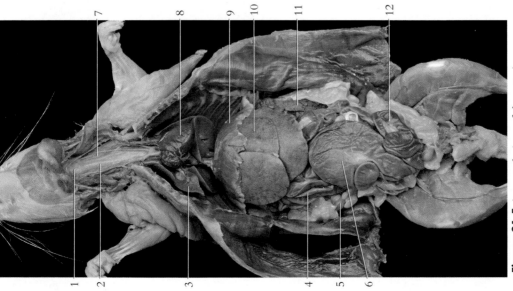

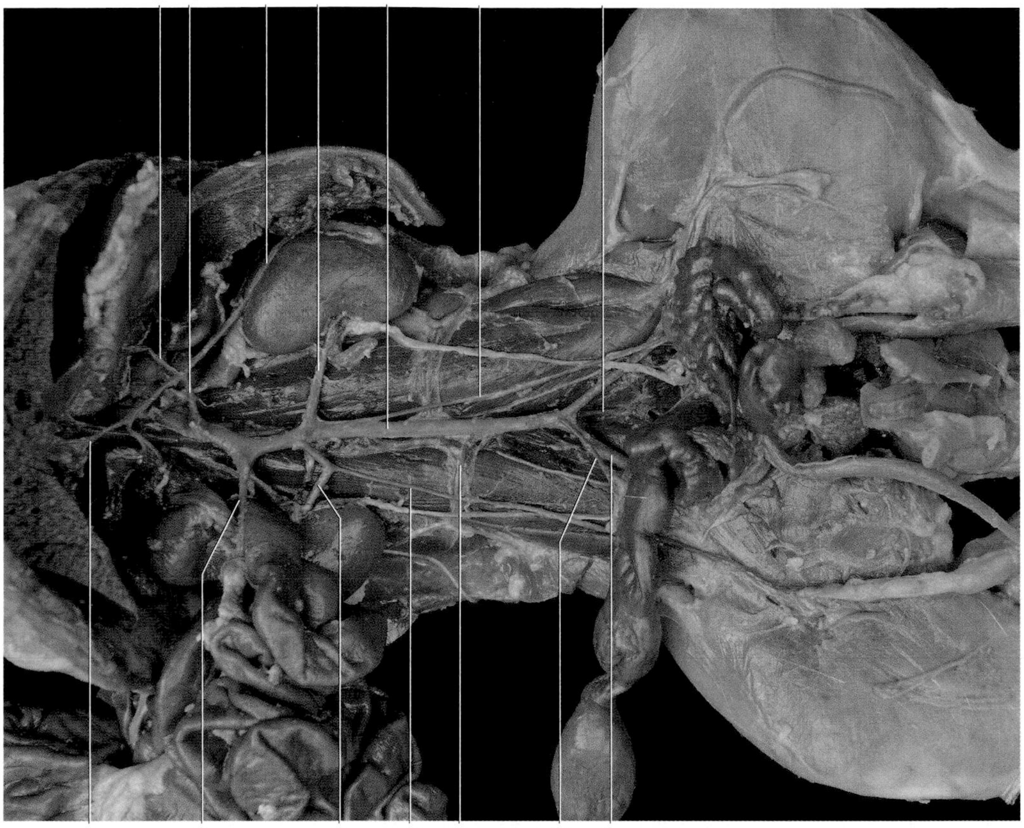

Figure 21.6 The abdominal arteries of the rat.

1. Hepatic artery
2. Right renal artery
3. Cranial mesenteric artery
4. Right testicular artery
5. Right iliolumbar artery
6. Caudal mesenteric artery (cut)
7. Right common iliac artery
8. Gastric artery
9. Celiac trunk
10. Splenic artery
11. Left renal artery
12. Abdominal aorta
13. Left testicular artery
14. Middle sacral artery

Figure 21.5 A ventral view of the rat viscera.

1. Larynx
2. Trachea
3. Right lung
4. Jejunum
5. Right ovary
6. Cecum
7. Esophagus
8. Heart
9. Diaphragm
10. Liver
11. Spleen
12. Ileum

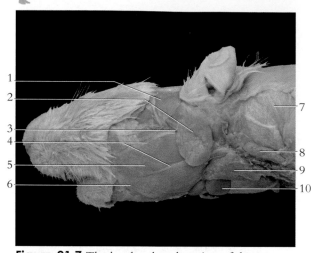

Figure 21.7 The head and neck region of the rat.

1. Temporalis m.
2. Extraorbital lacrimal gland
3. Extraorbital lacrimal duct
4. Facial nerve
5. Masseter m.
6. Parotid duct
7. Cervical trapezius m.
8. Parotid gland
9. Mandibular gland
10. Mandibular lymph node

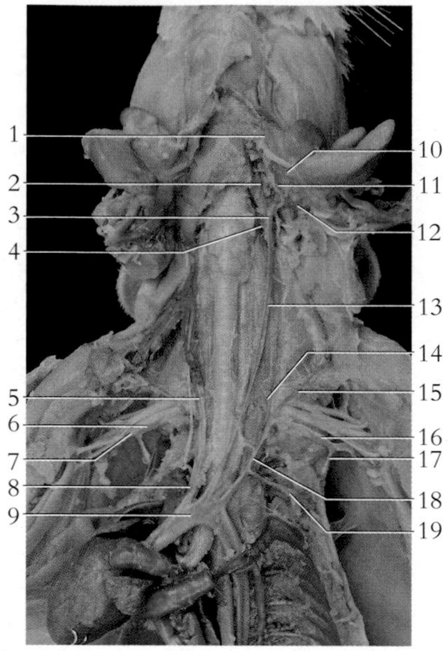

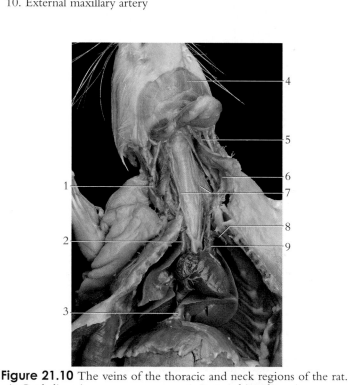

Figure 21.8 The arteries of the thoracic and neck regions of the rat.

1. Facial artery
2. Lingual artery
3. External carotid artery
4. Cranial thyroid artery
5. Right common carotid artery
6. Right axillary artery
7. Right brachial artery
8. Brachiocephalic artery
9. Aortic arch
10. External maxillary artery
11. Internal carotid artery
12. Occipital artery
13. Left common carotid artery
14. Vertebral artery
15. Cervical trunk
16. Lateral thoracic artery
17. Left axillary artery
18. Left subclavian artery
19. Internal thoracic artery

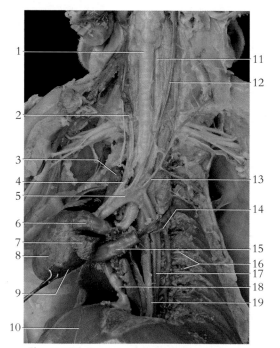

Figure 21.9 The rat heart (reflected) showing the major veins and arteries.

1. Trachea
2. Right common carotid artery
3. Right cranial vena cava (cut)
4. Brachiocephalic trunk
5. Aortic arch
6. Pulmonary trunk
7. Left auricle
8. Left ventricle
9. Coronary vein
10. Diaphragm
11. Esophagus
12. Left common carotid artery
13. Left subclavian artery
14. Azygos vein
15. Coronary sinus (cut)
16. Intercostal artery and vein
17. Aorta
18. Caudal vena cava
19. Esophagus

Figure 21.10 The veins of the thoracic and neck regions of the rat.

1. Cephalic vein
2. Right cranial vena cava
3. Caudal vena cava
4. Linguofacial vein
5. Maxillary vein
6. External jugular vein
7. Internal jugular vein
8. Lateral thoracic vein
9. Left cranial vena cava

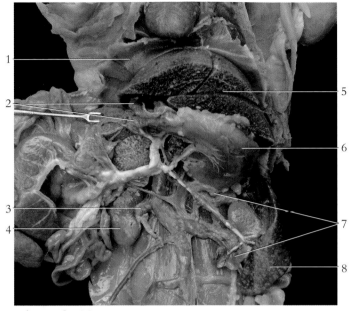

Figure 21.11 The abdominal viscera and vessels of the rat.
1. Diaphragm
2. Biliary and duodenal parts of pancreas
3. Right renal vein
4. Right kidney
5. Liver (cut)
6. Stomach
7. Gastrosplenic part of pancreas
8. Spleen

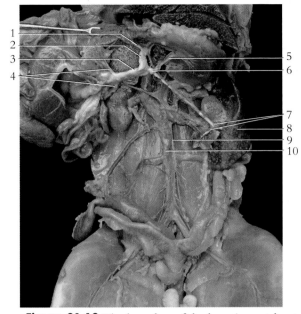

Figure 21.12 The branches of the hepatic portal system.
1. Cranial pancreaticoduodenal vein
2. Hepatic portal vein
3. Cranial mesenteric vein
4. Intestinal branches
5. Gastric vein
6. Gastrosplenic vein
7. Splenic branches
8. Spleen
9. Abdominal aorta
10. Caudal vena cava

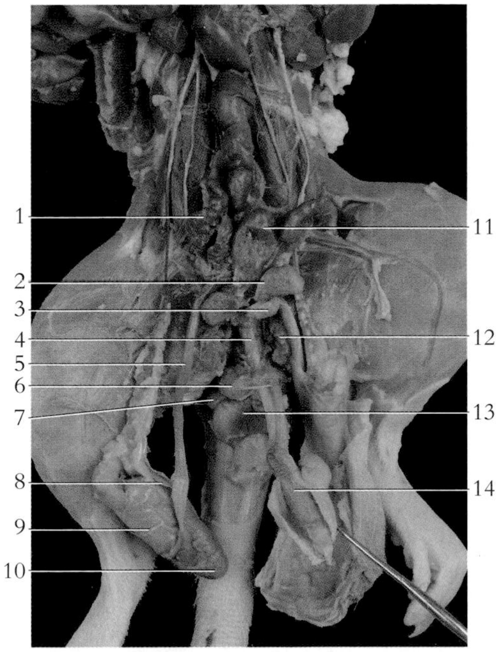

Figure 21.13 The urogenital system of the male rat.
1. Vesicular gland
2. Prostate (ventral part)
3. Prostate (dorsolateral part)
4. Urethra in the pelvic canal
5. Vas (ductus) deferens
6. Crus of penis
7. Bulbourethral gland
8. Head of epididymis
9. Testis
10. Tail of epididymis
11. Urinary bladder
12. Symphysis pubis (cut exposing pelvic canal)
13. Bulbocavernosus muscle
14. Penis

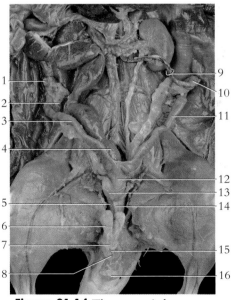

Figure 21.14 The urogenital system of the female rat.
1. Ovary
2. Uterine artery and vein
3. Uterine horn
4. Colon
5. Vagina
6. Preputial gland
7. Clitoris
8. Vaginal opening
9. Uterine artery and vein
10. Ovary
11. Uterine horn
12. Uterine body
13. Urinary bladder
14. Urethra
15. Urethral opening
16. Anus

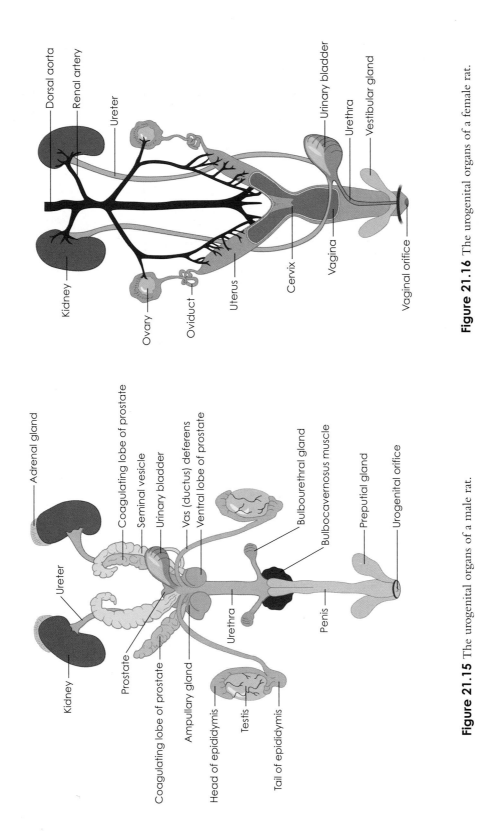

Figure 21.16 The urogenital organs of a female rat.

Figure 21.15 The urogenital organs of a male rat.

Glossary of Prefixes & Suffixes

Element	Definition and Example	Element	Definition and Example	Element	Definition and Example
a-	absent, deficient or without: atrophy	-cis	cut, kill: excision	gastro-	stomach: gastrointestinal
ab-	off, away from: abduct	co-	together: copulation	-gen	an agent that produces or originates: pathogen
abdomin-	abdomen	coel-	hollow cavity: coelom		
-able	capable of: viable	con-	with, together: congenital	-genic	produced from, producing: carcinogenic
ac-	toward, to: actin	contra-	against, opposite: contraception		
acou-	hear: acoustic	corn-	denoting hardness: cornified	gloss-	tongue: glossopharyngeal
ad-	denoting to, toward: adduct	corp-	body: corpus	glyco-	sugar: glycosuria
af-	movement toward a central point: afferent artery	crypt-	hidden: cryptorchism	-gram	a record, recording: myogram
		cyan-	blue color: cyanosis	gran-	grain, particle: agranulocyte
alba-	pale or white: linea alba	cysti-	sac or bladder: cystoscope		
-alg	pain: neuralgia			-graph	instrument for recording: electrocardiograph
ambi-	both: ambidextrous	cyto-	cell: cytology		
angi-	pertaining to vessel: angiology			grav-	heavy: gravid
ante-	before: antebrachium	de-	down, from: descent	gyn-	female sex: gynocology
anti-	against: anticoagulant	derm-	skin: dermatology		
aqua-	water: aqueous	di-	two: diarthrotic	hema(o)-	blood: hematology
archi-	to be first: archeteron	dipl-	double: diploid	haplo-	simple or single: haploid
arthri-	joint: arthritis	dis-	apart, away from: disarticulate	hemi-	half: hemiplegia
aud-	pertaining to ear: auditory			hepat-	liver: hepatic portal
		duct-	lead, conduct: ductus deferens	hetero-	other, different: heterosexual
auto-	self: autolysis	dur-	hard: dura mater		
		-dynia	pain: gastrodynia	histo-	tissue: histology
bi-	two: bipedal	dys-	bad, difficult, painful: dysentery	holo-	whole, entire: holocrine
bio-	life: biology			homo-	same, alike: homologous
blast-	generative or germ bud: osteoblast	e-	out, from: eccrine	hydro-	water: hydrocoel
		ecto-	outside, outer, external: ectoderm	hyper-	beyond, above, excessive: hypertension
brachi-	arm: brachialis				
brachy-	short: brachydont	-ectomy	surgical removal: tonsillectomy	hypo-	under, below: hypoglycemia
brady-	slow: bradycardia	ede-	swelling: edema		
bucc-	cheek: buccal cavity	-emia	pertaining to a condition of the blood: lipemia	-ia	state or condition: hypoglycemia
cac-	bad, ill: cachexia	end-	within: endoderm	-iatrics	medical specialties: pediatrics
calci-	stone: calculus			idio-	self, separate, distinct: idiopathic
capit-	head: capitis	entero-	intestine: enteritis	ilio-	ilium: iliosacral
carcin-	cancer: carcinogenic	epi-	upon, in addition: epidermis	infra-	beneath: infraspinatus
cardi-	heart: cardiac			inter-	among, between: intervertebral
caud-	tail: cauda equina	erythro-	red: erythrocyte	intra-	inside, within: intracellular
cata-	lower, under, against: catabolism	ex-	out of: excise		
-coel	swelling, and enlarged	exo-	outside: exocrine	-ion	process: acromion
		extra-	outside of, beyond, in addition: extracellular	iso-	equal, like: isotonic
				-ism	condition or state: rheumatism
	space or cavity: blastocoele	fasci-	band: fascia	-itis	inflammation: meningitis
cephal-	head: cephalis	febr-	fever: febrile		
cerebro-	brain: cerebrospinal fluid	-ferent	bear, carry: efferent arteriole	labi-	lip: labium majus
chol-	bile: cholic			lacri-	tears: nasolacrimal
chondr-	cartilage: chondrocyte	fiss-	split: fissure	later-	side: lateral
chrom-	color: chromocyte	for-	opening: foramen	-logy	science of: morphology
-cid(e)	destroy: germicide	-form	shape: fusiform	-lysis	solution, dissolve: hemolysis
circum-	around: circumduct			macro-	large, great: macrophage
				mal-	bad, abnormal, disorder: malignant

Element	Definition and Example	Element	Definition and Example	Element	Definition and Example
medi-	middle: medial	phag-	to eat: phagocyte	sub-	under, beneath, below: subcutaneous
mega-	great, large: megakaryocyte	-phil	have an affinity for: neutrophil	super-	above, beyond, upper: superficial
meso-	middle or moderate: mesoderm	phlebo-	vein: phlebitis	supra-	above, over: suprarenal
meta-	after, beyond: metatarsal	-phobe	abnormal fear, dread: hydrophobia	syn (sym)	together, joined, with: synapse
micro-	small: microtome	-plasty	reconstruction of: rhinoplasty	tachy-	swift, rapid: tachometer
mito-	thread: mitosis	platy-	flat, side: platysma	tele-	far: telencephalon
mono-	alone, one, single: monocyte	-plegia	stroke, paralysis: paraplegia	tens-	stretch: tensor fascia lata
mons-	mountain: mons pubis	-pnea	to breathe: apnea	tetra-	four: tetrad
morph-	form, shape: morphology	pneumato-	breathing: pneumonia	therm-	heat: thermogram
multi-	many, much: multinuclear	pod-	foot: podiatry	thorac-	chest: thoracic cavity
myo-	muscle: myology	-poieis	formation of: hematopoiesis	thrombo-	lump, clot: thrombocyte
		poly-	many, much: polyploid	-tomy	cut: appendectomy
narc-	numbness, stupor: narcotic	post-	after, behind: postnatal	tox-	poison: toxic
neo-	new, young: neonatal	pre-	before in time or place: prenatal	tract-	draw, drag: traction
necro-	corpse, dead: necrosis	prim-	first: primitive	trans-	across, over: transfuse
nephro-	kidney: nephritis	pro-	before in time or place: prosect	tri-	three: trigone
neuro-	nerve: neurolemma	proct-	anus: proctology	trich-	hair: trichology
noto-	back: notochord	pseudo-	false: pseudostratified	-trophy	a state relating to nutrition: hypertrophy
ob-	against, toward, in front of: obturator	psycho-	mental: psychology	-tropic	turning toward, changing: gonadotropic
oc-	against: occlusion	pyo-	pus: pyoculture		
-oid	resembling, likeness: sigmoid	quad-	fourfold: quadriceps femoris	ultra-	beyond, excess: ultrasonic
oligo-	few, small: oligodendrocyte	re-	back, again: repolarization	uni-	one: unicellular
-oma	tumor: lymphoma	rect-	straight: rectus abdominis	uro-	urine, urinary organs or tract: uroscope
oo-	egg: oocyte	reno-	kidney: renal	-uria	urine: polyuria
or-	mouth: oral	rete-	network: retina		
orchi-	testicles: cryptorchidism	retro-	backward: retroperitoneal	vas-	vessel: vasoconstriction
-ory	pertaining to: sensory	rhin-	nose: rhinitis	vermi-	worm: vermiform
osteo-	bone: osteoblast	-rrhage	excessive flow: hemorrhage	viscer-	organ: visceral
-ose	full of: adipose	-rrhea	flow, or discharge: diarrhea	vit-	life: vitamin
oto	ear: otolith			zoo-	animal: zoology
ovo-	egg: ovum	sanguin-	blood: sanguiferous	zygo-	union, join: zygote
para-	give birth to, bear: parturition	sarc-	flesh: sarcoplasm		
para-	near, beyond, beside: paranasal	-scope	instrument for examination of a part: stethoscope		
path-	disease, that which undergoes sickness: pathology	-sect	cut: dissect		
-pathy	abnormality, disease: neuropathy	semi-	half: semilunar		
ped-	children: pediatrician	serrate-	saw-edged: serratus anterior		
pen-	need, lack: penicillin	-stasis	standing: homeostasis		
-penia	deficiency: thrombocytopenia	-sis	state or condition: dialysis		
per-	through: percutaneous	steno-	narrow: stenohaline		
peri-	near, around: pericardium	-stomy	surgical opening: tracheotomy		

Glossary

A

abdomen (ab-do´men): the portion of the trunk located between the diaphragm and the pelvis; contains the abdominal cavity and its visceral organs.

abduction (ab-duk´shun): a movement away from the axis or midline of the body; opposite of adduction; a movement of a digit away from the axis of a limb.

acapnia (ah-kap´ne-ah): a decrease in normal amount of CO_2 in the blood.

accommodation (ah-kom-o-da´shun): a change in the shape of the lens of the eye so that vision is more acute; the focusing for various distances.

acetone (as´e-tone): an organic compound that may be present in the urine of diabetics; also called ketone bodies.

Achilles tendon (ah-kil´ez): see *tendo calcaneus*.

acidosis (as-i-do´sis): a disorder of body chemistry in which the alkaline substances of the blood are reduced below normal.

actin (ak´tin): a protein in muscle fibers that together with myosin is responsible for contraction.

acoustic (ah-koos´tik): referring to sound or the sense of hearing.

adduction (ah-duk´shun): a movement toward the axis or midline of the body; opposite of abduction; a movement of a digit toward the axis of a limb.

adenohypophysis (ad´e-no-hi-pof´i-sis): anterior pituitary gland.

adenoid (ad´e-noid): paired lymphoid structures in the naso-pharynx; also called pharyngeal tonsils.

adenosine triphosphate (ATP) (ah-den´o-sen tri-fos´fate): a chemical compound that provides energy for cellular use.

adipose (ad´e-pose): fat, or fat-containing, such as adipose tissue.

adrenal glands (ah-dre´nal): endocrine glands; one superior to each kidney; also called *suprarenal glands*.

aerobic (a-er-o´bik): requiring free O_2 for growth and metabolism as in the case of certain bacteria called *aerobes*.

allantois (ah-lan´to-is): an extraembryonic membranous sac that forms blood cells and gives rise to the fetal umbilical arteries and vein. It also contributes to the formation of the urinary bladder.

alveolus (al-ve´o-lus): an individual air capsule within the lung. Alveoli are the basic functional units of respiration. Also, the socket that secures a tooth.

amnion (am´ne-on): a membrane that surrounds the fetus to contain the amniotic fluid.

amphiarthrosis (am´fe-ar-thro´sis): a slightly moveable joint in a functional classification of joints.

anatomical position (an´ah-tom´e-kal): an erect body stance with the eyes directed forward, the arms at the sides, and the palms of the hands facing forward.

anatomy (ah-nat´o-me): the branch of science concerned with the structure of the body and the relationship of its organs.

antebrachium (an´te-bra´ke-um): the forearm.

anterior (ventral) (an-te´re-or): toward the front; the opposite of *posterior (dorsal)*.

antigen (an´te-jen): a substance which causes cells to produce antibodies.

anus (a´nus): the terminal end of the GI tract, opening of the anal canal.

aorta (a-or´tah): the major systemic vessel of the arterial portion of the circulatory system, emerging from the left ventricle.

apocrine gland (ap´o-krin): a type of sweat gland that functions in evaporative cooling.

appendix (ah-pen´diks): a short pouch that attaches to the cecum.

aqueous humor (a´kwe-us hu´mor): the watery fluid that fills the anterior and posterior chambers of the eye.

arachnoid mater (ah-rak´noid ma´ter): the weblike middle covering (meninx) of the central nervous system.

arbor vitae (ar´bor vi´tah): the branching arrangement of white matter within the cerebellum.

areola (ah-re´o-lah): the pigmented ring around the nipple.

artery (ar´ter-e): a blood vessel that carries blood away from the heart.

articular cartilage (ar-tik´u-lar kar´ti-lij): a hyaline cartilaginous covering over the articulating surface of bones of synovial joints.

ascending colon (ko´lon): the portion of the large intestine between the cecum and the right colic (hepatic) flexure.

atom (at´om): the smallest unit of an element that can exist and still have the properties of the element; collectively, atoms form molecules in a compound.

atrium (a´tre-um): either of two superior chambers of the heart that receive venous blood.

atrophy (at´ro-fe): a wasting away or decrease in size of a cell or organ.

auditory tube (aw´di-to´re): a narrow canal that connects the middle ear chamber to the pharynx; also called the *eustachian canal*.

autonomic (aw-to-nom´ik): self-governing; pertaining to the division of the nervous system which controls involuntary activities.

axilla (ak-sil´ah): the depressed hollow under the arm; the armpit.

axon (ak´son): The elongated process of a neuron (nerve cell) that transmits an impulse away from the cell body.

B

basement membrane: a thin sheet of extracellular substance to which the basal surfaces of membranous epithelial cells are attached.

basophil (ba´so-fil): a granular leukocyte that readily stains with basophilic dye.

belly: the thickest circumference of a skeletal muscle.

benign: (be-nine´): nonmalignant; a confined tumor.

blastula (blas´tu-lah): an early stage of prenatal development between the morula and embryonic stages.

blood: the fluid connective tissue that circulates through the cardiovascular system to transport substances throughout the body.

bolus (bo´lus): a moistened mass of food that is swallowed from the oral cavity into the pharynx.

bone: an organ composed of solid, rigid connective tissue, forming a component of the skeletal system.

Bowman's capsule (bo´manz kap´sul): see *glomerular capsule.*

brain: the enlarged superior portion of the central nervous system, located in the cranial cavity of the skull.

brain stem: the portion of the brain consisting of the medulla oblongata, pons, and midbrain.

bronchial tree (brong´ke-al): the bronchi and their branching bronchioles.

bronchiole (brong´ke-ol): a small division of a bronchus within the lung.

bronchus (bron´kus): a branch of the trachea that leads to a lung.

buccal cavity (buk´al): the mouth, or oral cavity.

bursa (ber´sah): a saclike structure filled with synovial fluid, which occurs around joints.

buttock (but´ok): the rump or fleshy mass on the posterior aspect of the lower trunk, formed primarily by the gluteal muscles.

C

calorie (kal´o-re): the unit of heat required to raise the temperature of one gram of water one degree centigrade.

calyx (ka´liks): a cup-shaped portion of the renal pelvis that encircles a renal papilla.

cancellous bone (kan´se-lus): spongy bone; bone tissue with a latticelike structure.

capillary (kap´i-lar´e): a microscopic blood vessel that connects an arteriole and a venule; the functional unit of the circulatory system.

carcinogenic (kar-si-no-jen´ik): stimulating or causing the growth of a malignant tumor, or cancer.

carpus (kar´pus): the proximal portion of the hand that contains the eight carpal bones.

cartilage (kar´ti´lij): a type of connective tissue with a solid elastic matrix.

caudal (kaw´dal): referring to a position more toward the tail.

cecum (se´kum): the pouchlike portion of the large intestine to which the ileum of the small intestine is attached.

cell: the structural and functional unit of an organism; the smallest structure capable of performing all the functions necessary for life.

central nervous system (CNS): the brain and the spinal cord.

centrosome (sen´tro-som): a dense body near the nucleus of a cell that contains a pair of centrioles.

cerebellum (ser´e-bel´um): the portion of the brain concerned with the coordination of movements and equilibrium.

cerebrospinal fluid (ser´e-bro-spi´nal): a fluid that buoys and cushions the central nervous system.

cerebrum (ser´e-brum): the largest portion of the brain, composed of the right and left hemispheres.

cervical (ser´vi-kal): pertaining to the neck or a necklike portion of an organ.

choanae (ko-a´na): the two posterior openings from the nasal cavity into the nasopharynx.

cholesterol (ko-les´ter-ol): an organic fatlike compound found in animal fat, bile, blood, liver, and other parts of the body.

chondrocyte (kon´dro-site): a cartilage cell.

chorion (ko´re-on): An extraembryonic membrane that participates in the formation of the placenta.

choroid (ko´roid): the vascular, pigmented middle layer of the wall of the eye.

chromosome (kro´mo-som): structure in the nucleus that contains the genes for genetic expression.

chyme (kime): The mass of partially digested food that passes from the stomach into the duodenum of the small intestine.

cilia (sil´e-ah): microscopic, hairlike processes that move in a wavelike manner on the exposed surfaces of certain epithelial cells.

ciliary body (sil´e-er´e): a portion of the choroid layer of the eye that secretes aqueous humor and contains the ciliary muscle.

circumduction (ser´kum-duk´shun): a conelike movement of a body part, such that the distal end moves in a circle while the proximal portion remains relatively stable.

clitoris (kli´to-ris): a small, erectile structure in the vulva of the female.

cochlea (kok´le-ah): the spiral portion of the inner ear that contains the spiral organ (organ of Corti).

ceolom (se´lom): the abdominal cavity.

colon (ko´lon): the first portion of the large intestine.

common bile duct: a tube that is formed by the union of the hepatic duct and cystic duct, transports bile to the duodenum.

compact bone: tightly packed bone that is superficial to spongy bone; also called *dense bone.*

condyle (kon´dile): a rounded process at the end of a long bone that forms an articulation.

connective tissue: one of the four basic tissue types within the body. It is a binding and supportive tissue with abundant matrix.

cornea (kor´ne-ah): the transparent, convex, anterior portion of the outer layer of the eye.

cortex (kor´teks): the outer layer of an organ such as the convoluted cerebrum, adrenal gland, or kidney.

costal cartilage (kos´tal): the cartilage that connects the ribs to the sternum.

cranial (kra´ne-al): pertaining to the cranium.

cranial nerve: one of twelve pairs of nerves that arise from the inferior surface of the brain.

D

dentin (den´tine): the principal substance of a tooth, covered by enamel over the crown and by cementum on the root.

dermis (der´mis): the second, or deep, layer of skin beneath the epidermis.

descending colon: the segment of the large intestine that descends on the left side from the level of the spleen to the level of the left iliac crest.

diaphragm (di´ah-fram): a flat dome of muscle and connective tissue that separates the thoracic and abdominal cavities.

diaphysis (di-af´i-sis): the shaft of a long bone.

diastole (di-as´to-le): the sequence of the cardiac cycle during which the ventricular heart chamber wall is relaxed.

diarthrosis (di´ar-thro´sis): a freely movable joint.

distal (dis´tal): away from the midline or origin; the opposite of *proximal*.

dorsal (dor´sal): pertaining to the back or posterior portion of a body part; the opposite of *ventral*.

ductus deferens (duk´tus def´er-enz): a tube that carries spermatozoa from the epididymis to the ejaculatory duct: also called the *vas deferens* or *seminal duct*.

duodenum (du´o-num): the first portion of the small intestine.

dura mater (du´rah ma´ter): the outermost meninx covering the central nervous system.

E

eccrine gland (ek´rin): a sweat gland that functions in body cooling.

ectoderm (ek´to-derm): the outermost of the three primary embryonic germ layers.

edema (e-de´mah): an excessive retention of fluid in the body tissues.

effector (ef-fek´tor): an organ such as a gland or muscle that responds to motor stimulation.

efferent (ef´er-ent): conveying away from the center of an organ or structure.

ejaculation (e-jak´u-la´shun): the discharge of semen from the male urethra during climax.

electrocardiogram (e-lek´tro-kar´de-o-gram´): a recording of the electrical activity that accompanies the cardiac cycle; also called ECG or EKG.

electroencephalogram (e-lek´tro-en-sef´ah-lo-gram): a recording of the brain wave pattern; also called EEG.

electromyogram (e-lek´tro-mi´o-gram): a recording of the activity of a muscle during contraction: also called EMG.

electrolyte (e-lek´tro-lite): a solution that conducts electricity by means of charged ions.

electron (e-lek´tron): the unit of negative electricity.

enamel (en-am´el): the outer, dense substance covering the crown of a tooth.

endocardium (en´do-kar´de-um): the fibrous lining of the heart chambers and valves.

endochondral bone (en´do-kon´dral): bones that form as hyaline cartilage models first and then are ossified.

endocrine gland (en´do-krine): hormone producing gland that is part of the endocrine system.

endoderm (en´do-derm): the innermost of the three primary germ layers of an embryo.

endometrium (en´do-me´tre-um): the inner lining of the uterus.

endothelium (en´do-the´le-um): the layer of epithelial tissue that forms the thin inner lining of blood vessels.

eosinophil (e´o-sin´o-fil): a type of white blood cell that becomes stained by acidic eosin dye; constitutes about 2%–4% of the white blood cells.

epicardium (ep´i-kar´de-um): the thin, outer layer of the heart: also called the *visceral pericardium*.

epidermis (ep´i-der´mis): the outermost layer of the skin, composed of stratified squamous epithelium.

epididymis (ep´i-did´i-mis): a coiled tube located along the posterior border of the testis; stores spermatozoa and discharges them during ejaculation.

epidural space (ep´i-du´ral): a space between the spinal dura mater and the bone of the vertebral canal.

epiglottis (ep´i-glot´is): a leaflike structure positioned on top of the larynx that covers the glottis during swallowing.

epinephrine (ep´i-nef´rin): a hormone secreted from the adrenal medulla resulting in action similar to those from sympathetic nervous system stimulation; also called *adrenaline*.

epiphyseal plate (ep´i-fize-al): a cartilaginous layer located between the epiphysis and diaphysis of a long bone and functions in longitudinal bone growth.

epiphysis (e-pif´i-sis): the end segment of a long bone, distinct in early life but later becoming part of the larger bone.

epithelial tissue (ep´i-the´le-al): one of the four basic tissue types; the type of tissue that covers or lines all exposed body surfaces.

erythrocyte (e-rith´ro-site): a red blood cell.

esophagus (e-sof´ah-gus): a tubular organ of the GI tract that leads from the pharynx to the stomach.

estrogen (es´tro-jen): female sex hormone secreted from the ovarian (Graafian) follicle.

eustachian canal (u-sta´ke-an): see *auditory tube*.

excretion (eks-kre´shun): discharging waste material.

exocrine gland (ek´so-krin): a gland that secretes its product to an epithelial surface, directly or through ducts.

expiration (ek´spi-ra´shun): the process of expelling air from the lungs through breathing out; also called *exhalation.*

extension (ek-sten´shun): a movement that increases the angle between two bones of a joint; opposite of flexion.

external ear: the outer portion of the ear, consisting of the auricle (pinna), and external auditory canal.

extracellular (esk-trah-sel´u-lar): outside a cell or cells.

extrinsic (eks-trin´sik): pertaining to an outside or external origin.

F

facet (fas´et): a small, smooth surface of a bone where articulation occurs.

fallopian tube (fal-lo´pe-an): see *uterine tube.*

fascia (fash´e-ah): a tough sheet of fibrous connective tissue binding the skin to underlying muscles or supporting and separating muscle.

fasciculus (fah-sik´u-lus): a bundle of muscle or nerve fibers.

feces (fe´sez): waste material expelled from the GI tract during defecation, composed of food residue, bacteria, and secretions; also called *stool.*

fetus (fe´tus): the unborn offspring during the last stage of prenatal development.

filtration (fil-tra´shun): the passage of a liquid through a filter or a membrane.

fimbriae (fim´bre-e): fringelike extensions from the borders of the open end of the uterine tube.

fissure (fish´ure): a groove or narrow cleft that separates two parts of an organ.

flexion (flek´shun): a movement that decreases the angle between two bones of a joint; opposite of extension.

fontanel (fon´tah-nel): a membranous-covered region on the skull of a fetus or baby where ossification has not yet occurred: also called a *soft spot.*

foot: the terminal portion of the lower extremity, consisting of the tarsus, metatarsus, and digits.

foramen (fo-ra´men): an opening in an anatomical structure for the passage of a blood vessel or a nerve.

foramen ovale (o-val´e): the opening through the interatrial septum of the fetal heart.

fossa (fos´ah): a depressed area, usually on a bone.

fourth ventricle (ven´tri-k´l): a cavity within the brain containing cerebrospinal fluid.

fovea centralis (fo´ve-ah sen´tra´lis): a depression on the macula lutea of the eye where only cones are located, which is the area of keenest vision.

G

gallbladder: a pouchlike organ, attached to the inferior side of the liver, which stores and concentrates bile.

gamete (gam´ete): a haploid sex cell, sperm or egg.

gamma globulins (gam´mah glob´u-lins): protein substances often found in immune serums that act as antibodies.

ganglion (gang´gle-on): an aggregation of nerve cell bodies outside the central nervous system.

gastrointestinal tract (gas´tro-in-tes´tin-al): the tubular portion of the digestive system that includes the stomach and the small and large intestines; also called the *GI tract.*

gene (jene): one of the biologic units of heredity; parts of the DNA molecule located in a definite position on a certain chromosome.

genetics (je-net´iks): the study of heredity.

gingiva (jin-ji´vah): the fleshy covering over the mandible and maxilla through which the teeth protrude within the mouth; also called the *gum.*

gland: an organ that produces a specific substance or secretion.

glans penis (glanz pe´nis): the enlarged, distal end of the penis.

glomerular capsule (glo-mer´u-lar): the double-walled proximal portion of a renal tubule that encloses the glomerulus of a *nephron;* also called *Bowman's capsule.*

glomerulus (glo-mer´u´lus): a coiled tuft of capillaries that is surrounded by the glomerular capsule and filters urine from the blood.

glottis (glot´is): a slitlike opening into the larynx, positioned between the vocal folds.

glycogen (gli´ko-jen): the principal storage carbohydrate in animals. It is stored primarily in the liver and is made available as glucose when needed by the body cells.

goblet cell: a unicellular gland within columnar epithelia that secretes mucus.

gonad (go´nad): a reproductive organ, testis or ovary, that produces gametes and sex hormones.

gray matter: the portion of the central nervous system that is composed of nonmyelinated nervous tissue.

greater omentum (o-men´tum): a double-layered peritoneal membrane that originates on the greater curvature of the stomach and extends over the abdominal viscera.

gut: pertaining to the intestine; generally a developmental term.

gyrus (ji´rus): a convoluted elevation or ridge.

H

hair: an epidermal structure consisting of keratinized dead cells that have been pushed up from a dividing basal layer.

hair cells: specialized receptor nerve endings for responding to sensations, such as in the spiral organ of the inner ear.

hair follicle (fol´li-k´l): a tubular depression in the skin in which a hair develops.

hand: the terminal portion of the upper extremity, consisting of the carpus, metacarpus, and digits.

hard palate (pal´at): the bony partition between the oral and nasal cavities, formed by the maxillae and palatine bones.

haustra (haws´trh): sacculations or pouches of the colon.

haversian system (ha-ver´shan): see *osteon.*

heart: a muscular, pumping organ positioned in the thoracic cavity.

hematocrit (he-mat´o-krit): the volume percentage of red blood cells in whole blood.

hemoglobin (he´mo-glo´bin): the pigment of red blood cells that transports O_2 and CO_2.

hemopoiesis (hem´ah-poi-e´sis): production of red blood cells.

hepatic portal circulation (por´tal): the return of venous blood from the digestive organs and spleen through a capillary network within the liver before draining into the heart.

heredity (he-red´i-te): the transmission of certain characteristics, or traits, from parents to offspring, via the genes.

hiatus (hi-a´tus): an opening or fissure.

hilum (hi´lum): a concave or depressed area where vessels or nerves enter or exit an organ.

histology (his-tol´o-je): microscopic anatomy of the structure and function of tissues.

homeostasis (ho-me-o-sta´sis): a consistency and uniformity of the internal body environment which maintains normal body function.

hormone (hor´mone): a chemical substance that is produced in an endocrine gland and secreted into the bloodstream to cause an effect in a specific target organ.

hyaline cartilage (hi´ah-line): the most common kind of cartilage in the body, occurring at the articular ends of bones, in the trachea, and within the nose, and forms the precursor to most of the bones of the skeleton.

hymen (hi´men): a developmental remnant (vestige) of membranous tissue that partially covers the vaginal opening.

hyperextension (hi´per-ek-sten´shun): extension beyond the normal anatomical position of 180°.

hypothalamus (hi´po-thal´ah-mus): a structure within the brain below the thalamus, which functions as an autonomic center and regulates the pituitary gland.

I

ileocecal valve (il´e-o-se´kal): a specialization of the mucosa at the junction of the small and large intestine that forms a one-way passage and prevents the backflow of food materials.

ileum (il´e-um): the terminal portion of the small intestine between the jejunum and cecum.

inferior vena cava (ve´nah ka´vah): a systemic vein that collects blood from the body regions inferior to the level of the heart and returns it to the right atrium.

inguinal (ing´gwi-nal): pertaining to the groin region.

inguinal canal: the passage in the abdominal wall through which a testis descends into the scrotum.

insertion: the more moveable attachment of a muscle, usually more distal in location.

inspiration (in´spi-ra´shun): the act of breathing air into the alveoli of the lungs; also called *inhalation.*

integument (in-teg´u-ment): pertaining to the skin.

internal ear: the innermost portion or chamber of the ear, containing the cochlea and the vestibular organs.

interstitial (in-ter-stish´al): pertaining to spaces or structures between the functioning active tissue of any organ.

intracellular (in-trah-sel´u-lar): within the cell itself.

intervertebral disc (in´ter-ver´te-bral): a pad of fibrocartilage between the bodies of adjacent vertebrae.

intestinal gland (in-tes´ti-nal): a simple tubular digestive gland that opens onto the surface of the intestinal mucosa and secretes digestive enzymes; also called *crypt of Lieberkühn.*

intrinsic (in-trin´sik): situated in or pertaining to an internal organ.

iris (i´ris): the pigmented, vascular tunic portion of the eye that surrounds the pupil and regulates its diameter.

islets of Langerhans (i´lets of lahng´er-hanz): see *pancreatic islets.*

isotope (i´so-tope): a chemical element that has the same atomic number as another but a different atomic weight.

isthmus (is´mus): a narrow neck or portion of tissue connecting two structures.

J

jejunum (je-joo´num): the middle portion of the small intestine, located between the duodenum and the ileum.

joint capsule (kap´sule): the fibrous tissue that encloses the joint cavity of a synovial joint.

jugular (jug´u-lar): pertaining to the veins of the neck which drain the areas supplied by the carotid arteries.

K

karyotype (kar´e-o-tip): the arrangement of chromosomes that is characteristic of the species or of a certain individual.

keratin (ker´ah-tin): an insoluble protein present in the epidermis and in epidermal derivatives such as hair and nails.

kidney (kid´ne): one of the paired organs of the urinary system that contains nephrons and filters wastes from the blood in the formation of urine.

L

labia majora (la´be-ah ma-jor´ah): a portion of the external genitalia of a female, consisting of two longitudinal folds of skin extending downward and backward from the mons pubis.

labia minora (ma-nor´ah): two small folds of skin, devoid of hair and sweat glands, lying between the labia majora of the external genitalia of a female.

lacrimal gland (lak´ri-mal): a tear-secreting gland, located on the superior lateral portion of the eyeball underneath the upper eyelid.

lactation (lak-ta´shun): the production and secretion of milk by the mammary glands.

lacteal (lak´te-al): a small lymphatic duct within a villus of the small intestine.

lacuna (lah-ku´nah): a hollow chamber that houses an osteocyte in mature bone tissue or a chondrocyte in cartilage tissue.

lamella (lah-mel´ah): a concentric ring of matrix surrounding the central canal in an osteon of mature bone tissue.

large intestine: the last major portion of the GI tract, consisting of the cecum, colon, rectum, and anal canal.

larynx (lar´inks): the structure located between the pharynx and trachea that houses the vocal folds (cords); commonly called the *voice box.*

lens (lenz): a transparent refractive structure of the eye, derived from ectoderm and positioned posterior to the pupil and iris.

leukocyte (lu´ko-site): a white blood cell; also spelled *leucocyte.*

ligament (lig´ah-ment): a fibrous band or cord of connective tissue that binds bone to bone to strengthen and provide support to the joint; also may support viscera.

limbic system (lim´bik): a portion of the brain concerned with emotions and autonomic activity.

linea alba (lin´e-ah al´bah): a fibrous band extending down the anterior medial portion of the abdominal wall.

locus (lo´kus): the specific location or site of a gene within the chromosome.

lumbar (lum´bar): pertaining to the region of the loins.

lumen (lu´men): the space within a tubular structure through which a substance passes.

lung: one of the two major organs of respiration within the thoracic cavity.

lymph (limf): a clear fluid that flows through lymphatic vessels.

lymph node: a small, ovoid mass located along the course of lymph vessels.

lymphocyte (lim´fo-site): a type of white blood cell.

M

macula lutea (mak´u-lah lu´te-ah): a depression in the retina that contains the fovea centralis, the area of keenest vision.

malignant (mah-lig´nant): a disorder that becomes worse and eventually causes death, as in cancer.

malnutrition (mal-nu-trish´un): any abnormal assimilation of food; receiving insufficient nutrients.

mammary gland (mam´er-e): the gland of the female breast responsible for lactation and nourishment of the young.

marrow (mar´o): the soft vascular tissue that occupies the inner cavity of certain bones and produces blood cells.

matrix (ma´triks): the intercellular substance of a tissue.

meatus (me-a´tus): an opening or passageway into a structure.

mediastinum (me´de-as-ti´num): the partition in the center of the thorax between the two pleural cavities.

medulla (me-dul´ah): the center portion of an organ.

medulla oblongata (ob´long-ga´tah): a portion of the brain stem between the pons and the spinal cord.

medullary cavity (med´u-lar´e): the hollow center of the diaphysis of a long bone, occupied by marrow.

meiosis (mi-o´sis): cell division by which gametes, or haploid sex cells, are formed.

melanocyte (mel´ah-no-site): a pigment-producing cell in the deepest epidermal layer of the skin.

membranous bone (mem´brah-nus): bone that forms from membranous connective tissue rather than from cartilage.

menarche (me-nar´ke): the first menstrual discharge.

meninges (me-nin´jez): a group of three fibrous membranes that cover the central nervous system.

menisci (men-is´si): wedge-shaped cartilages in certain synovial joints.

menopause (men´o-pawz): the cessation of menstrual periods in the human female.

menses (men´sez): the monthly flow of blood from the female genital tract.

menstrual cycle (men´stru-al): the rhythmic female reproductive cycle, characterized by changes in hormone levels and physical changes in the uterine lining.

menstruation (men´stru-a´shun): the discharge of blood and tissue from the uterus at the end of the menstrual cycle.

mesentery (mes´en-ter´e): a fold of peritoneal membrane that attaches an abdominal organ to the abdominal wall.

mesoderm (mes´o-derm): the middle one of the three primary germ layers.

mesothelium (mes´o-the´leum): a simple squamous epithelial tissue that lines body cavities and covers visceral organs; also called *serosa.*

metabolism (me-tab´o-lizm): the chemical changes that occur within a cell.

metacarpus (met´ah-kar´pus): the region of the hand between the wrist and the digits, including the five bones that support the palm of the hand.

metastasis (me-tas´tah-sis): the spread of a disease from one organ or body part to another.

metatarsus (met´ah-tar´sus): the region of the foot between the ankle and the digits, containing five bones.

microbiology (mi-kro-bi-ol´o-je): the science dealing with microscopic organisms, including bacteria, fungi, viruses, and protozoa.

microvilli (mi´kro-vil´i): microscopic, invaginations of cell membranes on certain epithelial cells.

midbrain: the portion of the brain between the pons and the forebrain.

middle ear: the middle of the three ear chambers, containing the three auditory ossicles.

mitosis (mi-to´sis): the process of cell division, in which the two daughter cells are identical and contain the same number of chromosomes.

mitral valve (mi´tral): the left atrioventricular heart valve; also called the bicuspid valve.

mixed nerve: a nerve containing both motor and sensory nerve fibers.

molecule (mol'e-kule): a minute mass of matter, composed of a combination of atoms that form a given chemical substance or compound.

motor neuron (nu'ron): a nerve cell that conducts action potential away from the central nervous system and innervates effector organs (muscles and glands); also called *efferent neuron*.

motor unit: a single motor neuron and the muscle fibers it innervates.

mucosa (mu-ko'sah): a mucous membrane that lines cavities and tracts opening to the exterior.

muscle (mus'cl): an organ adapted to contract; three types of muscle tissue are cardiac, smooth, and skeletal.

mutation (mu-ta'shun): a variation in an inheritable characteristic, a permanent transmissible change in which the offspring differ from the parents.

myelin (mi'e-lin): a lipoprotein material that forms a sheathlike covering around nerve fibers.

myocardium (mi'o-kar'de-um): the cardiac muscle layer of the heart.

myofibril (mi'o-fi'bril): a bundle of contractile fibers within muscle cells.

myoneural junction (mi'o-nu'ral): the site of contact between an axon of a motor neuron and a muscle fiber.

myosin (mi'o-sin): a thick filament protein that together with actin causes muscle contraction.

N

nail: a hardened, keratinized plate that develops from the epidermis and forms a protective covering on the dorsal surfaces of the digits.

nares (na'rez): the opening into the nasal cavity; also called *nostrils*.

nasal cavity (na'zal): a mucosa-lined space above the oral cavity, which is divided by a nasal septum and is the first chamber of the respiratory system.

nasal septum (sep'tum): a bony and cartilaginous partition that separates the nasal cavity into two portions.

nephron (nef'ron): the functional unit of the kidney, consisting of a glomerulus, glomerular capsule, convoluted tubules, and the loop of the nephron.

nerve: a bundle of nerve fibers outside the central nervous system.

neurofibril node (nu'ro-fi'bril): a gap in the myelin sheath of a nerve fiber; also called the *node of Ranvier*.

neuroglia (nu-rog'le-ah): specialized supportive cells of the central nervous system.

neurolemmocyte (nu'ri-lem-o'site): a specialized neuroglia cell that surrounds an axon fiber of a peripheral neuron and forms the neurilemmal sheath; also called the *Schwann cell*.

neuron (nu'ron): the structural and functional unit of the nervous system, composed of a cell body, dendrites, and an axon; also called a *nerve cell*.

neutrophil (nu'tro-fil): a type of phagocytic white blood cell.

nipple: a dark pigmented, rounded projection at the tip of the breast.

node of Ranvier (rah-ve-a'): see *neurofibril node*.

notochord (no'to-kord): a flexible rod of tissue that extends the length of the back of an embryo.

nucleus (nu'kle-us): a spheroid body within a cell that contains the genetic factors of the cell.

nurse cells: specialized cells within the testes that supply nutrients to developing spermatozoa; also called *sertoli cells* or *sustentacular cells*.

O

olfactory (ol-fak'to-re): pertaining to the sense of smell.

oocyte (o'o-site): a developing egg cell.

oogenesis (o'o-hen'e-sis): the process of female gamete formation.

optic (op'tik): pertaining to the eye and the sense of vision.

optic chiasma (ki-as'mah): an X-shaped structure on the inferior aspect of the brain where there is a partial crossing over of fibers in the optic nerves.

optic disc: a small region of the retina where the fibers of the ganglion neurons exit from the eyeball to form the optic nerve; also called the *blind spot*.

oral: pertaining to the mouth; also called *buccal*.

organ: a structure consisting of two or more tissues, which performs a specific function.

organelle (or'gan-el'): a minute structure of a cell with a specific function.

organism (or'gah-nizm): an individual living creature.

orifice (or'i fis): an opening into a body cavity or tube.

origin (or'i-jin): the place of muscle attachment onto the more stationary point or proximal bone; opposite the insertion.

osmosis (os-mo'sis): the passage of a solvent, such as water, from a solution of lesser concentration to one of greater concentration through a semipermeable membrane.

ossicle (os'si-kel): one of the three bones of the middle ear.

osteocyte (os'te-o-site): a mature bone cell.

osteon (os'te-on): a group of osteocytes and concentric lamellae surrounding a central canal within bone tissue; also called a *haversian system*.

ovarian follicle (o-va're-an fol'li-k'l): a developing ovum and its surrounding epithelial cells.

ovary (o'vah-re): the female gonad in which ova and certain sexual hormones are produced.

oviduct (o'vi-dukt): the tube that transports ova from the ovary to the uterus; also called the *uterine tube* or *fallopian tube*.

ovulation (o'vu-la'shun): the rupture of an ovarian follicle with the release of an ovum.

ovum (o'vum): a secondary oocyte after ovulation but before fertilization.

P

palate (pal´at): the roof of the oral cavity.

palmar (pal´mar): pertaining to the palm of the hand.

pancreas (pah´kre-as): organ in the abdominal cavity that secretes gastric juices into the GI tract and insulin and glucagon into the blood.

pancreatic islets (pan´kre-at´ik): a cluster of cells within the pancreas that forms the endocrine portion of the pancreas; also called *islets of Langerhans.*

papillae (pah-pil´e): small fingerlike projections.

paranasal sinus (par´ah-na´zal si´nus): an air chamber lined with a mucous membrane that communicates with the nasal cavity.

parasympathetic (par´ah-smi´pah-thet´ik): pertaining to the division of the autonomic nervous system concerned with activities that are antagonistic to sympathetic.

parathyroids (par´ah-thi´roids): small endocrine glands that are embedded on the posterior surface of the thyroid glands and are concerned with calcium metabolism.

parietal (pah-ri´e-tal): pertaining to a wall of an organ or cavity.

parotid gland: (pah-rot´id): one of the paired salivary glands on the side of the face over the masseter muscle.

parturition (par´tu-rish´un): the process of childbirth.

pathogen (path´o-jen): any disease-producing organism.

pectoral girdle (pek´to-ral): the portion of the skeleton that supports the upper extremities.

pelvic (pel´vik): pertaining to the pelvis.

pelvic girdle: the portion of the skeleton to which the lower extremities are attached.

penis (pe´nis): the external male genital organ, through which urine passes during urination and which transports semen to the female during coitus.

pericardium (per´i-kar´de-um): a protective serous membrane that surrounds the heart.

perineum (per´i-ne´um): the floor of the pelvis, which is the region between the anus and the scrotum in the male and between the anus and the vulva in the female.

periosteum (per´e-os´te-um): a fibrous connective tissue covering the outer surface of bone.

peripheral nervous system (pe-rif´er-al): the nerves and ganglia of the nervous system that lie outside of the brain and spinal cord.

peristalsis (per´i-stal´sis): rhythmic contractions of smooth muscle in the walls of various tubular organs, which move the contents along.

peritoneum (per´i-to-ne´um): the serous membrane that lines the abdominal cavity and covers the abdominal viscera.

phagocyte (fag´o-site): any cell that engulfs other cells, including bacteria, or small foreign particles.

phalanx (fa´lanks), pl. *phalanges*: a bone of the finger or toe.

pharynx (far´inks): the organ of the GI tract and respiratory system located at the back of the oral and nasal cavities and extending to the larynx anteriorly and the esophagus posteriorly; also called the *throat.*

physiology (fiz´e-ol´o-je): the science that deals with the study of body functions.

pia mater (pi´ah ma´ter): the innermost meninx that is in direct contact with the brain and spinal cord.

pineal gland (pin´e-al): a small cone-shaped gland located in the roof of the third ventricle.

pituitary gland (pi-tu´i-tar´e): a small, pea-shaped endocrine gland situated on the inferior surface of the brain that secretes a number of hormones; also called the *hypophysis* and commonly called the *"master gland."*

placenta (plah-sen´tah): the organ of metabolic exchange between the mother and the fetus.

plasma (plaz´mah): the fluid, extracellular portion of circulating blood.

platelets (plate´lets): fragments of specific bone marrow cells that function in blood coagulation: also called *thrombocytes.*

pleural membranes (ploor´al): serous membranes that surround the lungs and line the thoracic cavity.

plexus (plek´sus): a network of interlaced nerves or vessels.

plica circulares (pli´kah ser-ku-lar´is): a circular deep fold within the wall of the small intestine that increases the absorptive surface area.

pons (ponz): the portion of the brain stem just above the medulla oblongata and anterior to the cerebellum.

posterior (dorsal) (pos-ter´e-or): toward the back.

pregnancy: a condition where a female has a developing offspring in the uterus.

prenatal (pre-na´tal): the period of offspring development during pregnancy; before birth.

proprioceptor (pro´pre-o-sep´tor): a sensory nerve ending that responds to changes in tension in a muscle or tendon.

prostate (pros´tate): a walnut-shaped gland surrounding the male urethra just below the urinary bladder that secretes an additive to seminal fluid during ejaculation.

proximal (prok´si-mal): closer to the midline of the body or origin of an appendage: opposite of *distal.*

puberty (pu´ber-te): the period of development in which the reproductive organs become functional.

pulmonary (pul´mo-ner´e): pertaining to the lungs.

pupil: the opening through the iris that permits light to enter the vitreous chamber of the eyeball and be refracted by the lens.

R

receptor (re-sep´tor): a sense organ or a specialized end of a sensory neuron that receives stimuli from the environment.

rectum (rek´tum): the portion of the GI tract between the sigmoid colon and the anal canal.

reflex arc: the basic conduction pathway through the nervous

system, consisting of a sensory neuron, interneuron, and a motor neuron.

renal (re′nal): pertaining to the kidney.

renal corpuscle (kor′pus′l): the portion of the nephron consisting of the glomerulus and a glomerular capsule.

renal pelvis: the inner cavity of the kidney formed by the expanded ureter and into which the calyces open.

replication (re-pli-ka′shun): the process of producing a duplicate; a copying or duplication, such as DNA replication.

respiration: (res′pi-ra′shun): the exchange of gases between the external environment and the cells of an organism.

rete testis (re′te tes′tis): a network of ducts in the center of the testis, site of spermatozoa transport.

retina (ret′i-nah): the inner layer of the eye that contains the rods and cones.

retraction (re-trak′shun): the movement of a body part, such as the mandible, backward on a plane parallel with the ground; the opposite of *protraction.*

rod: a photoreceptor in the retina of the eye that is specialized for colorless, dim light vision.

rotation (ro-ta′shun): the movement of a bone around its own longitudinal axis.

rugae (ru′je): the folds or ridges of the mucosa of an organ.

S

sagittal (saj′i-tal): a vertical plane through the body that divides it into right and left portions.

salivary gland (sal′i-ver-e): an accessory digestive gland that secretes saliva into the oral cavity.

sarcolemma (sar′ko-lem′ah): the cell membrane of a muscle fiber.

sarcomere (sar′ko-mere): the portion of a skeletal muscle fiber between the two adjacent Z lines that is considered the functional unit of a myofibril.

Schwann cell (shwahn): see *neurolemmocyte.*

sclera (skle′rah): the outer white layer of connective tissue that forms the protective covering of the eye.

scrotum (skro′tum): a pouch of skin that contains the testes and their accessory organs.

sebaceous gland (se-ba′shus): an exocrine gland of the skin that secretes *sebum,* an oily protective product.

semen (se′men): the secretion of the reproductive organs of the male, consisting of spermatozoa and additives.

semicircular ducts: tubular channels within the inner ear that contain the receptors for equilibrium; also called *semicircular canals.*

semilunar valve (sem′e-lu′nar): crescent-shaped heart valves, positioned at the entrances to the aorta and the pulmonary trunk.

seminal vesicles (sem′i-nal ves′i-k′lz): a pair of accessory male reproductive organs lying posterior and inferior to the urinary bladder, which secrete additives to spermatozoa into the ejaculatory ducts.

sensory neuron (nu′ron): a nerve cell that conducts an impulse from a receptor organ to the central nervous system; also called *afferent neuron.*

serous membrane (se′rus): an epithelial and connective tissue membrane that lines body cavities and covers viscera; also called *serosa.*

sesamoid bone (ses′ah-moid): a membranous bone formed in a tendon in response to joint stress.

sigmoid colon (sig′moid ko′lon): the S-shaped portion of the large intestine between the descending colon and the rectum.

sinoatrial node (sin′o-a′tre-al): a mass of specialized cardiac tissue in the wall of the right atrium that initiates the cardiac cycle; the SA node; also called the *pacemaker.*

sinus (si′nus): a cavity or hollow space within a body organ such as a bone.

skeletal muscle: a type of muscle tissue that is multinucleated, occurs in bundles, has crossbands of proteins, and contracts either in a voluntary or involuntary fashion.

small intestine: the portion of the GI tract between the stomach and the cecum, functions in absorption of food nutrients.

smooth muscle: a type of muscle tissue that is nonstriated, composed of fusiform, single-nucleated fibers, and contracts in an involuntary, rhythmic fashion within the walls of visceral organs.

somatic (so-mat′ik): pertaining to the nonvisceral parts of the body.

spermatic cord (sper′mat′ik): the structure of the male reproductive system composed of the ductus deferens, spermatic vessels, nerve, cremasteric muscle, and connective tissue.

spermatogenesis (sper′mah-to-jen′e-sis): the production of male sex gametes, or spermatozoa.

spermatozon (sper′mah-to-zo′on): a sperm cell, or gamete.

sphincter (sfingk′ter): a circular muscle that constricts a body opening or the lumen of a tubular structure.

spinal cord (spi′nal): the portion of the central nervous system that extends from the brain stem through the vertebral canal.

spinal nerve: one of the thirty-one pairs of nerves that arise from the spinal cord.

spleen: a large, blood-filled organ located in the upper left of the abdomen and attached by the mesenteries to the stomach.

spongy bone (spun′je): a type of bone that contains many porous spaces; also called *cancellous bone.*

stomach: a pouchlike digestive organ between the esophagus and the duodenum.

submucosa (sub′mu-ko′sah): a layer of supportive connective tissue that underlies a mucous membrane.

superior vena cava (ve′nah ka′vah): a large systemic vein that collects blood from regions of the body superior to the heart and returns it to the right atrium.

surfactant (ser-fak´tant): a substance produced by the lungs that decreases the surface tension within the alveoli.

suture (su´chur): a type of fibrous joint articulating between bones of the skull.

sympathetic (sim´pah-thet´ik): pertaining to that part of the autonomic nervous system concerned with activities antagonistic to the parasympathetic.

synapse (sin´aps): a minute space between the axon terminal of a presynaptic neuron and a dendrite of a postsynaptic neuron.

synovial cavity (si-no´ve-al): a space between the two bones of a synovial joint, filled with synovial fluid.

system: a group of body organs that function together.

systole (sis´to-le): the muscular contraction of the ventricles of the heart during the cardiac cycle.

systolic pressure (sis´tol´ik): arterial blood pressure during the ventricular systolic phase of the cardiac cycle.

T

target organ: the specific body organ that a particular hormone affects.

tarsus (tahr´sus): pertaining to the ankle; the proximal portion of the foot that contains the seven tarsal bones.

tendo calcaneous (ten´do kal-ka´ne-us): the tendon that attaches the calf muscles to the calcaneous bone.

tendon (ten´dun): a band of dense regular connective tissue that attaches muscle to bone.

testis (tes´tis): the primary reproductive organ of a male, which produces spermatozoa and male sex hormones.

thoracic (tho-ras´ik): pertaining to the chest region.

thoracic duct: the major lymphatic vessel of the body, which drains lymph from the entire body except the upper right quadrant and returns it to the left subclavian vein.

thorax (tho´raks): the chest.

thymus gland (thi´mus): a bi-lobed lymphoid organ positioned in the upper mediastinum, posterior to the sternum and between the lungs.

tissue: an aggregation of similar cells and their binding intercellular substance, joined to perform a specific function.

tongue: a protrusible muscular organ on the floor of the oral cavity.

trachea (tra´ke-ah): the airway leading from the larynx to the bronchi; also called the *windpipe*.

tract: a bundle of nerve fibers within the central nervous system.

transverse colon (ko´lon): a portion of the large intestine that extends from right to left across the abdomen between the hepatic and splenic flexures.

tricuspid valve (tri-kus´pid): the heart valve between the right atrium and the right ventricle.

tympanic membrane (tim-pan´ik): the membranous eardrum positioned between the outer and middle ear; also called the *tympanum*, or the *ear drum*.

U

umbilical cord (um-bil´i-kal): a cordlike structure containing the umbilical arteries and vein, which connects the fetus with the placenta.

umbilicus (um-bil´i-kus): the site where the umbilical cord was attached to the fetus: also called the *navel*.

ureter (u-re´ter): a tube that transports urine from the kidney to the urinary bladder.

urethra (u-re´thrah): a tube that transports urine from the urinary bladder to the outside of the body.

urinary bladder (u´re-ner´e): a distensible sac in the pelvic cavity which stores urine.

uterine tube (u´ter-in): the tube through which the ovum is transported to the uterus and where fertilization takes place: also called the *oviduct* or *fallopian tube*.

uterus (u´ter-us): a hollow, muscular organ in which a fetus develops. It is located within the female pelvis between the urinary bladder and the rectum.

uvula (u´vu-lah): a fleshy, pendulous portion of the soft palate that blocks the nasopharynx during swallowing.

V

vagina (vah-ji´nah): a tubular organ that leads from the uterus to the vestibule of the female reproductive tract and receives the male penis during coitus.

vein: a blood vessel that conveys blood toward the heart.

ventral (ven´tral): toward the front surface of the body: also called *anterior*.

vestibular folds: the supporting folds of tissue for the vocal folds within the larynx.

vestibular window: a membrane-covered opening in the bony wall between the middle and inner ear, into which the footplate of the stapes fits; also called *oval window*.

viscera (vis´er-ah): the organs within the abdominal or thoracic cavities.

vitreous humor (vit´re-us hu´mor): the transparent gel that occupies the space between the lens and retina of the eye.

vocal folds: folds of the mucous membrane in the larynx that produce sound as they are pulled taut and vibrated; also called *vocal cords*.

vulva (vul´vah): the external genitalia of the female that surround the opening of the vagina; also called the *pudendum*.

Z

zygote (zi´gote): a fertilized egg cell formed by the union of a sperm and an ovum.

Index